AF838210

SCIENTIFIC NETWORKING AND THE GLOBAL HEALTH NETWORK SUPERCOURSE

NATO Security through Science Series

This Series presents the results of scientific meetings supported under the NATO Programme for Security through Science (STS).

Meetings supported by the NATO STS Programme are in security-related priority areas of Defence Against Terrorism or Countering Other Threats to Security. The types of meeting supported are generally "Advanced Study Institutes" and "Advanced Research Workshops". The NATO STS Series collects together the results of these meetings. The meetings are co-organized by scientists from NATO countries and scientists from NATO's "Partner" or "Mediterranean Dialogue" countries. The observations and recommendations made at the meetings, as well as the contents of the volumes in the Series, reflect those of participants and contributors only; they should not necessarily be regarded as reflecting NATO views or policy.

Advanced Study Institutes (ASI) are high-level tutorial courses to convey the latest developments in a subject to an advanced-level audience

Advanced Research Workshops (ARW) are expert meetings where an intense but informal exchange of views at the frontiers of a subject aims at identifying directions for future action

Following a transformation of the programme in 2004 the Series has been re-named and re-organised. Recent volumes on topics not related to security, which result from meetings supported under the programme earlier, may be found in the NATO Science Series.

The Series is published by IOS Press, Amsterdam, and Springer Science and Business Media, Dordrecht, in conjunction with the NATO Public Diplomacy Division.

Sub-Series

A.	Chemistry and Biology	Springer Science and Business Media
B.	Physics and Biophysics	Springer Science and Business Media
C.	Environmental Security	Springer Science and Business Media
D.	Information and Communication Security	IOS Press
E.	Human and Societal Dynamics	IOS Press

http://www.nato.int/science
http://www.springeronline.nl
http://www.iospress.nl

Sub-Series D: Information and Communication Security – Vol. 5 ISSN: 1574-5589

Scientific Networking and the Global Health Network Supercourse

Edited by

Faina Linkov

and

Ronald LaPorte

University of Pittsburgh Graduate School of Public Health, Pittsburgh, PA, USA

IOS Press

Amsterdam • Berlin • Oxford • Tokyo • Washington, DC

Published in cooperation with NATO Public Diplomacy Division

Proceedings of the NATO Advanced Research Workshop on
Scientific Networking and the Global Health Network Supercourse for the Prevention of
Man Made and Natural Disasters
Kaunas, Lithuania
8–11 August 2005

© 2006 IOS Press.

All rights reserved. No part of this book may be reproduced, stored in a retrieval system,
or transmitted, in any form or by any means, without prior written permission from the publisher.

ISBN 1-58603-631-9
Library of Congress Control Number: 2006929179

Publisher
IOS Press
Nieuwe Hemweg 6B
1013 BG Amsterdam
Netherlands
fax: +31 20 687 0019
e-mail: order@iospress.nl

Distributor in the UK and Ireland
Gazelle Books Services Ltd.
White Cross Mills
Hightown
Lancaster LA1 4XS
United Kingdom
fax: +44 1524 63232
e-mail: sales@gazellebooks.co.uk

Distributor in the USA and Canada
IOS Press, Inc.
4502 Rachael Manor Drive
Fairfax, VA 22032
USA
fax: +1 703 323 3668
e-mail: iosbooks@iospress.com

LEGAL NOTICE
The publisher is not responsible for the use which might be made of the following information.

PRINTED IN THE NETHERLANDS

Scientific Networking and the Global Health Network Supercourse
F. Linkov and R. LaPorte (Eds.)
IOS Press, 2006
© 2006 IOS Press. All rights reserved.

Scientific Networking and the Global Health Network Supercourse: Introduction

Faina LINKOV, PhD
University of Pittsburgh

Natural and man-made disasters, like floods, eruptions of volcanoes, etc., have tormented mankind since antiquity. Despite all the warning signs, it is certain that when the eruption of Vesuvius started on the morning of 24 August, CE 79, it caught the local population of Pompeii utterly unprepared for the major disaster that ultimately ruined the entire city. What makes our world today different from the population of ancient Pompeii 2,000 years ago is better abilities to share scientific data about the warning signs of disasters. Modern technologies are giving us an unprecedented opportunity to share disaster preparedness and mitigation information very rapidly and effectively. Technologies such as the Internet, telecommunications, etc., have a great potential to help us prepare for disasters, however, oftentimes scientists are not taking advantage of them. In an effort to change this situation, our group conducted a NATO funded workshop on the scientific networking.

Networking of scientists from various disciplines is an important aspect of disaster forecast and prevention. Since the breakdown of the Soviet Union in 1991, man-made disasters became a very real possibility in Russia and other Former Soviet Union (FSU) countries. Collapse of public health infrastructure in the countries of the FSU became a realistic threat to the health of the people throughout the FSU, Europe, the Mediterranean region, and worldwide. An earthquake or famine in the FSU could be the tipping point leading to a chain reaction of political, economic, and health destabilization in the FSU, markedly escalating nuclear and other types of threat for all other countries. FSU scientists have few connections with their colleagues in Europe and the Western world, and their work is rarely published in peer-reviewed journals. It is likely that the improvement in scientific networking among FSU scientists and their counterparts in other countries would lead to better disaster information sharing and ultimately to better prevention and mitigation of natural and man-made disasters. FSU scientists networking with U.S. scientists and others worldwide would be a powerful form of public health mitigation against natural disasters, and escalating man-made disasters (e.g., nuclear war). The Global Health Network Supercourse (GHNS), supported initially by NASA, and then by the National Institutes of Health (NIH) created a network of over 600 scientists in the FSU region. As a part of the Supercourse project, directors of this workshop were successful in networking over 31,000 public health professionals in 151 countries. This network shares a free library of 2,500 low-bandwidth Supercourse lectures at www.pitt.edu/~super1. The Supercourse project is currently supported by the National Library of Medicine (NIH).

The networking of scientists researching the area of natural and man-made disasters became the topic of NATO funded Advanced Research Workshop that took place in Kaunas, Lithuania, in August 2005. Strengthening and improving existing networking and finding new cost-effective information distribution channels became the leading topic of this workshop. This workshop demonstrated how to use cost-effective Internet- based information distribution channels to:

- Share disaster low-bandwidth preparedness and mitigation research information in the library of the Supercourse lectures as demonstrated at <www.pitt.edu/~super1>
- Create scientific networking locally and with other countries
- Collaborate on disaster preparedness and mitigation research projects
- Increase one's chances of publication in peer-reviewed journals

This workshop was significant not only because it allowed exchanging the knowledge among scientists from NATO countries, but also because it provided a fertile ground for forging life-long friendships. These friendships will serve as a cornerstone for the future multinational collaborative projects in the area of disaster preparedness and mitigation. As meeting participants discussed in one of the sessions, networking is not about computers or other technologies, but it is about friends. Improved networking among friends in the area of disaster preparedness and mitigation is promising to make us better prepared for crisis situations in NATO countries and around the world.

Acknowledgments

Editors would like to thank all authors who contributed to this book for their high quality manuscripts. This work could not have been accomplished without generous support from North Atlantic Treaty Organization (NATO) and Department of Epidemiology, Graduate School of Public Health, University of Pittsburgh. We would also like to thank Kaunas University of Medicine and Vitae Litera, Kaunas, Lithuania, for their organizational and logistical support of the project. Finally, the editors would like to thank Janet Horsch, MLS, MFA, for her medical editing services.

Contents

Scientific Networking and the Global Health Network Supercourse

Scientific Networking and the Global Health Network Supercourse
F. Linkov and R. LaPorte (Eds.)
IOS Press, 2006
© 2006 IOS Press. All rights reserved.

Disasters: Introduction and State of the Art

Eric K. Noji
Centers for Disease Control and Prevention, Washington, DC

Original citation:
Eric Noji, Disasters: Introducation and State of the Art, Epidemiologic Reviews, 2005, Volume 27: 3-8, by permission of Oxford University Press and the Society for Epidemiologic Research.

Fifteen years have passed since the last update on this topic was published in Epidemiologic Reviews [1] and 24 years since the first [2]. In the intervening years, disaster prevention, mitigation, and preparedness have evolved in important ways [3]. Clearly, it was time to update the last review. Fifteen years ago, disaster management was simply left to a few dedicated professionals. Roles were clear: Rescue workers rushed to help victims, and certain agencies stepped in to provide temporary shelter and food. Usually within weeks after the disaster's impact, most people forgot about the disaster—until the next one came to wreak new destruction. Unfortunately, disasters throughout the world, such as the series of four destructive hurricanes that struck the southeast coast of the United States from August to September of 2004 [4] and the tsunami disaster in December 2004, have provided ample opportunities to test the policies and recommendations set out in the late 1980s. At least 80 percent of the population growth in the 1990s has occurred in towns and cities. According to the United Nations, in the year 2005, one half of the world's population will live in urban areas, crowded onto just 3 percent of the earth's land. This is an alarming increase in population density. Problems inherent in such rapid growth are especially unwieldy in developing countries; 17 of the 20 largest cities are now in developing countries compared with seven of 20 in 1950. By 2025, 80 percent of the world's population will reside in developing countries. One of every two large cities in the developing world is vulnerable to natural disasters such as floods, severe storms, and earthquakes [3].

Fortunately, over the past decade, the public health approach to disasters has changed significantly. Today, the management of humanitarian assistance involves many more and different players, and disaster management is recognized as a significant priority of the public health system. Today, prevention, mitigation, and preparedness are part of the vocabulary of public health officials in national and international organizations and, more importantly, they are used to advance the cause of reducing mortality and morbidity from disasters [5].

Epidemiology, as the applied instrument of public health interventions, can provide much needed information on which a rational, effective, and flexible policy for the management of disasters can be based. In particular, epidemiology provides the tools for rapid and effective problem solving during public health emergencies, such as natural and technologic disasters and emergencies from terrorism.

After sudden-impact disasters, time constraints and disruption of an area's infrastructure have frequently made it necessary to conduct rapid assessment surveys using nonprobability sampling methods. These methods may produce biased results because they are often based on purposive, convenience, or haphazard selection of subjects for interview [6]. In the last 15 years, investigators demonstrated the use of a modified cluster-sampling method to performa rapid needs assessment after hurricanes [7; 8]. In the first survey conducted 3 days after Hurricane Andrew struck south Florida in August 1992, clusters were systematically selected from a heavily damaged area by using a grid that had been overlaid on aerial photographs. Survey teams interviewed seven occupied households in consecutive order in each selected cluster. Results were available within 24 hours of beginning the survey. Surveys of the same heavily damaged area and of a less severely affected area were conducted 7 and 10 days later, respectively.

Initial survey workers found few households with injured residents, but a large proportion of households were without telephones or electricity. The workers' findings convinced disaster relief workers to focus on providing primary care and preventive services to residents rather than to divert resources in order to establish unnecessary mass-casualty trauma services. The cluster-survey method used in this rapid assessment was modified from methods developed by the World Health Organization's Expanded Programme on Immunization to assess vaccine coverage. Although cluster surveys have been used in refugee settings to assess nutritional and health status, this activity represented the first use of the Expanded Programme on Immunization survey Correspondence to Dr. Eric K. Noji, CDC Washington Office, 200 Independence Avenue, SW, Room 719-B, method to obtain population-based data after a suddenimpact natural disaster.

Although cluster-survey techniques hold promise for providing information rapidly after a disaster, in certain settings these techniques may be less applicable. For example, epidemiologists who used a cluster-survey technique after the January 1994 earthquake in Northridge, California, found that the technique needed modification. Unlike the damage from hurricanes, which is generally uniform over a large geographic area and thus can support the use of cluster sampling, earthquake-related damage varies considerably, with some areas experiencing little destruction and others experiencing heavy destruction. The extent of damage after earthquakes depends on local soil conditions, the distance and rate of ground-shaking attenuation from the epicenter, and the quality of building construction. Therefore, using a cluster-sampling approach to assess damages after an earthquake may cause health authorities to miss seriously affected areas and, thus, to underestimate overall damages [9].

Results of epidemiologic studies of disasters have not only led to the scientific measurement and description of disaster-associated health effects but also been used to identify groups in the population at particular risk for adverse health events, to help emergency managers match resources to needs, to monitor the effectiveness of relief efforts, to improve contingency planning, and to formulate recommendations for decreasing the adverse public health consequences of future disasters [3]. Unfortunately, it is assumed by many that all disaster research has been and will be based upon "scientific evidence." Evidence consists of data upon which a judgment or conclusion may be based.

Evidence must be "valid." Evidence-based disaster medicine, therefore, supposedly is based upon facts. For example, disaster planning is only as good as the assumptions on which it is based. However, these assumptions are often based on conventional wisdom and stereotypes rather than on systematically collected evidence. While these assumptions may be logical, what is logical is not always what is true [10]. For example, it is often assumed that disasters trigger widespread panic and leave stricken populations helpless and dependent on government authorities and rescue and relief organizations for strong leadership and assistance. Disaster planning often focuses on what these agencies can do for the public with the view that the public can do little for itself.

Planners may be unaware that there exists a large body of evidence on disaster responses that has been collected over several decades by rapid-response field teams from scientific institutions specializing in disaster research. This evidence shows that panic is extremely rare in disasters and that members of the public in the impact area will take the initiative to help themselves and others.

Most postearthquake or post-building collapse search and rescue, for example, is carried out not by police, firefighters, and formally trained rescue teams but rather by the survivors themselves (family members, neighbors, coworkers, friends, and those who just happen to be in the area) [11]. To the lay public, the best emergency care is seen as transport as quickly as possible to the closest hospital. If ambulances are not immediately available in sufficient numbers, the survivors will use whatever means of transport that is expedient to accomplish that objective (e.g., private car, bus, taxi, or even on foot). During the September 11, 2001, World Trade Center attack, for example, only 6.7 percent of the casualties were transported by ambulance.

As a result, in most disasters the closest hospitals receive most of the casualties, while those slightly farther away wait for casualties that never arrive. Furthermore, field triage, first aid, and decontamination stations are often bypassed because those transporting victims are unaware of their existence or location, or because they believe that better care is available at hospitals. This all happens very quickly, with hospitals usually receiving no warning that a disaster has even occurred and, most importantly, that they will be inundated with casualties beginning to arrive within a few minutes.

Officials who are unaware of this evidence may inadvertently create dysfunctional plans. For example, they may designate one hospital to receive casualties contaminated by hazardous substances. They may assume that the fire department will decontaminate casualties at the scene, or that hospital staff will have advanced notice so they can don chemical protective suits and set up decontamination equipment before patients arrive.

Evidence collected by epidemiologists is also useful for planning. For example, the primary focus for disaster medical planning has traditionally been on hospital treatment of the critically injured. However, evidence from epidemiologic studies indicates that most disaster injuries are relatively minor and could easily be treated in urgent care centers, private physicians' offices, outpatient surgery centers, and clinics—sparing hospitals for the more serious cases. Additional evidence suggests that many postdisaster visits to hospital emergency departments are for medical conditions other than injuries [10]. In some cases, these patients are elderly people who have lost access to their routine sources of medical care (e.g., pharmacies, private doctors, home health care). Yet, there seems to be little planning to ensure that these sources of medical care can survive, function, or expand capacity in disasters [4].

Evaluation of the medical and public health responses to disasters is one of the principal responsibilities of epidemiologists with an eye toward progressive improvement in the ability of the health system to respond more effectively and efficiently to disasters [2]. Such responses must be evaluated from the perspective of their outcomes and to what extent these interventions benefited victims of the disaster, especially relative to the goals that were expected by such planned responses. Epidemiologists have used a great variety of data collection methods and strategies to study the postdisaster health effects of major disasters involving acute events, such as earthquakes and tropical cyclones. Primarily using descriptive epidemiology, they have collected large amounts of epidemiologic data through case studies of new and previous disasters.

However, interventions also may be evaluated with regard to prevention or mitigation of the effect of an event. Such evaluations often are difficult, as their success is assessed by the fact that nothing happened that could have happened. Washington, DC 20201 (e-mail: exn1@cdc.gov). Identification of risk factors for death or injury will require a more sophisticated, analytical approach. Such analytical studies have usually been of a case-control design. For example, why did some people die while their neighbors, family members, or others survived? Isolated case studies of the relation between death or injuries and the type of traditional housing structures have provided clear indications regarding simple measures to be implemented in order to reduce human losses. Such analyses following disasters have yielded new information that has altered traditional thinking about the prevention of disaster-related mortality. Results of epidemiologic research on disasters have formed the scientific basis for increasingly effective prevention and intervention strategies to decrease mortality in several disaster situations. For example, epidemiologic studies of tornadoes have resulted in changes in local housing and land-use regulations regarding the danger of mobile homes and have formed the basis of National Weather Service safety guidelines issued to citizens in tornado-prone parts of the country [12]. Results of epidemiologic investigations of a wide spectrum of adverse medical and health consequences of disasters have allowed us to target specific interventions to prevent specific disaster-related health effects (e.g., improved warning and evacuation before flash floods and tropical cyclones [13], the identification of effective safety actions that building occupants should take during earthquakes [14], and the development of measures to avoid clean-up injuries following hurricanes [15]), to measure the effectiveness of disaster prevention and preparedness programs, and to help local communities develop better emergency preparedness and mitigation programs. More analytical studies such as these are needed to test conventional warnings and public safety advisories [5].

Despite the existence of useful, systematically collected evidence from hundreds of disasters, this body of knowledge can quickly become out-of-date [1,;2]. This is because of changes in disaster threats (e.g., pandemic influenza, suicide terrorism, specific targeting of medical personnel in war zones) and in the health-care system (advances in emergency medical service systems, emergency department overcrowding, nursing shortages, closures of trauma centers). In addition, there still exist critical data gaps in how the health-care system deals with disasters. An example is the lack of systematically collected data on the medical and public health response to releases of hazardous chemicals. Furthermore, we lack an effective, nationally institutionalized process of knowledge transfer for gathering and disseminating lessons learned from health and medical responses to disasters from researchers to first responders.

Fifteen years ago, the term ''complex humanitarian disasters'' was not commonly used. The focus of attention was usually the plight of refugees fleeing conflicts related to the tensions between the two superpowers, the Soviet Union and the United States. Much has changed in the intervening years. First, the geopolitical context has altered dramatically, with an initial increase in the intensity and scope of Cold War-related conflicts in the 1980s followed by the collapse of the Soviet Union and the subsequent ''epidemic'' of ethnic and religious conflicts. Second, the public health impact of armed conflicts on civilian populations has significantly worsened, especially during the years since the end of the Cold War. Third, on a more positive note, there has been a steady increase in technical publications in the form of journal articles, books, and manuals documenting public health outcomes and proposing more effective responses to conflict-associated population emergencies. The term complex humanitarian disaster reflects the multicausal nature and complicated response mechanisms of recent emergencies. In terms of their public health impact, complex humanitarian disasters may be defined as ''relatively acute situations affecting large populations, caused by a combination of factors, generally including civil strife or war, often exacerbated by food shortages and population displacement, and resulting in significant excess mortality'' [16, p. 1012].

The public health impact of complex disasters in the 1990s has been extensively documented. The Lancet and JAMA have both published reports on emergencies in northern Iraq, Somalia, Bosnia and Herzegovina, Nepal, and Zaire. One useful article that appeared following the Somali emergency documented the different approaches to the collection of public health information among various agencies [17]. Several years later, however, when 1 million Rwandan refugees fled into the eastern Zaire province of North Kivu, there was a remarkable degree of cooperation and standardization of information-gathering methods among the agencies present. This was reflected in a landmark article jointly authored by 24 epidemiologists from the Zaire Ministry of Health, World Health Organization, United Nations High Commissioner for Refugees, US Centers for Disease Control and Prevention, Me´decins Sans Frontie`res (Doctors without Borders), the French Army, and the Red Cross [18].

Major advances have been made during the past decade in the way the international community responds to the health and nutrition consequences of complex emergencies. The public health and clinical response to diseases of acute epidemic potential has improved, especially in camps. Casefatality rates for severely malnourished children have plummeted because of better protocols and products. Renewed focus is required on the major causes of death in conflict-affected societies—particularly, acute respiratory infections, diarrhea, malaria, measles, neonatal causes, and malnutrition—outside camps and often across regions and even political boundaries. In emergencies in sub-Saharan Africa, particularly, southern Africa, human immunodeficiency virus/acquired immunodeficiency syndrome is also an important cause of morbidity and mortality. Stronger coordination, increased accountability, and a more strategic positioning of nongovernmental organizations and United Nations agencies are crucial to achieving lower maternal and child morbidity and mortality rates in complex emergencies [19,;20].

While our understanding of the public health problems of refugees and displaced persons steadily improved, the causes of and response mechanisms to man-made emergencies became significantly more complicated. Whereas the focus of assistance programs in the 1970s and early 1980s was on refugees who had crossed borders to escape armed conflicts, in the 1990s it was often necessary to provide assistance to civilians still in the proximity of the conflict or displaced within their own countries. Civil wars in the Darfur region of western Sudan, Somalia, Liberia, Sierra Leone, Angola, Afghanistan, Chechnya, Sri Lanka, East Timor, and the former Yugoslavia had profound and tragic effects on the health of local civilian populations [21]. Today, complex emergencies are humanitarian crises that involve, if not war, then high levels of violence. Increasingly, civilians have become the intentional target of violence. Hundreds of thousands of civilians have been trapped in urban enclaves and siege-like situations where public utilities have been destroyed and basic medical services have collapsed. Children have been forcibly conscripted into opposing armed forces and have proved to be the most violent and pitiless of combatants. The provision of humanitarian assistance in these settings has proved extremely difficult and dangerous. The symbol of the Red Cross is no longer a guarantee of neutrality or even safety, with dozens of staff from the International Committee of the Red Cross murdered. Similar targeting and assassination of United Nations and humanitarian relief workers have now become an accepted hazard of doing such work. We all mourn the death of Sergio Viera de Mello, United Nations Special Representative to Iraq, on August 19, 2003.

Since Somalia in 1992, military involvement has often become essential for the provision of security, intelligence, and logistic support to international relief organizations. In fact, without such assistance in Bosnia, Kosovo, Liberia, Sierra Leone, and the recent tsunami disaster in South Asia, relief operations would have ground to a halt. As could be predicted, the decade of the 1990s brought much confusion and uncertainty to the traditional humanitarian relief community (both government and nongovernment) in the role of the military in complex emergencies. Usually, in these situations, organizations such as the United Nations (e.g., World Health Organization, United Nations Children's Fund, United Nations High Commissioner for Refugees), nongovernmental organizations, the International Committee of the Red Cross, and the International Federation of Red Cross have retained overall coordination, if not leadership and control. Until recently, this alone was felt to be essential for maintaining the neutrality (and thus safety) of relief workers. However, since the Balkan wars, Western militaries (particularly those of the United States) have substantially increased their activities in humanitarian projects, such as providing field clinical hospitals, water, sanitation, communicable disease control, food programs, and community health. Despite the humanitarian motives for such military operations, many relief organizations believe that this engagement contributes to the danger to their field staff by blurring the lines between civilian and military function and falsely associates them with the military forces. This became a major issue in the planning, execution, and recovery of the health infrastructure in Iraq following the end of major hostilities in Operation Iraqi Freedom. One author of World in Crisis concluded his chapter on the role of medical relief agencies as follows: ''many issues remain unresolved and hotly debated ... foremost is the challenge of working in the hostile and threatening environment of armed conflict in a global climate of political indecisiveness and moral inconsistency. While we await a concrete manifestation of the muchheralded new world order, relief agencies and the individuals who make up their field teams will continue to work on the front lines in an ethical limbo'' [22, p. 134].

On September 11, 2001, the United States experienced the worst terrorist attack in its history. As the nation sought to deal with this tragedy, it would face a second wave of terrorism—this time, in the form of a biologic attack. There should be no doubt by now that the challenge of terrorism has left an indelible mark on the world as we know it, spanning all inhabited continents, crossing all cultures, and penetrating the borders of all countries.

Unfortunately, a disaster caused by the intentional release of biologic weapons would be very different from other natural or technologic disasters, conventional military strikes, or even attacks with other weapons of mass destruction (e.g., nuclear, chemical, or explosive). The initial responders to a biologic disaster will most likely include county and city health officers, hospital staff, members of the outpatient medical community, and a wide range of personnel in the public health system and not traditional first responders such as police, fire, rescue, and ambulance services. Expanded public health laboratory capacity, increased surveillance (disease monitoring), early alert, warning and outbreak response capacity, and health communication and training are critical for an effective response to bioterrorism—with the focus of such public health preparedness resources and expertise at the state and local levels.

It is likely that recognition of the nature of and appropriate response to future bioterrorist attacks and natural epidemics, such as West Nile virus, pandemic influenza, and the international outbreak of severe acute respiratory syndrome (SARS), also will unfold over time. This is a difficult lesson in an age of 24-hour media coverage and expectations of instant answers. It is critical that public health authorities familiarize the communities they serve and the media with the likelihood that reliable answers to questions arising in future attacks will take time to assemble and validate [23]. Furthermore, the public must understand that messages (including medical advice, recommendations about who is at risk, and treatment) conveyed at one given point in time, although based on the best available information, are subject to change when new facts become known.

The myth that things go back to normal within a few weeks is especially pernicious. The truth is that the effects of a disaster last a long time [24]. Disaster-affected countries deplete many of their financial and material resources in the immediate postimpact phase. The bulk of the need for external assistance is in the restoration of normal primary health-care services, water systems, housing, and income-producing work. The longer-term recovery and rehabilitation needs in the affected areas are more poorly understood than the short-term needs, but they may be even more important. Many of the large relief agencies have substantial capacity for both relief and development, but effecting a transition from relief activities to sustainable and meaningful reconstruction activities is neither a simple nor a straightforward task. Relief organizations still have much to learn about shifting from short-term medical-aid efforts to productive, sustainable interventions that promote the development of a local health-care system [25].

In particular, social and mental health problems will appear when the acute crisis has subsided and the victims feel (and often are) abandoned to their own means [10]. Unfortunately, mental health and the psychological consequences of disasters have not yet received the attention they deserve in the epidemiologic literature (especially compared with the extensive research and body of knowledge in the social and behavioral sciences). Research of populations affected by disasters (whether due to civil conflict or volcanic eruptions) has revealed that these populations may have to cope with widespread depression, anxiety, and posttraumatic stress for years after the disaster [24]. Successful relief programs incorporate long-term mental health-care services in their overall planning for disaster relief, recovery, and reconstruction.

The physical health of survivors may also be adversely affected for years, particularly in technologic disasters, such as those involving chemicals and radiation. For example, the most pressing health concerns associated with nuclearreactor incidents are the delayed effects from long-term exposure to low levels of radiation [26]. Cancer or genetic defects induced by radiation exposure may not appear until many years after exposure and may be induced by low levels of exposure [26].

Epidemiologic study of disasters has clearly increased during the past 15 years, especially in university and academic communities, and it has undergone an explosive acceleration since September 11, 2001. Dozens of new professional societies and scientific forums for the presentation of original work in the epidemiology of disasters appeared. Dozens of university research centers now concentrate on the health and medical effects of disasters, including collaborating centers under the sponsorship of the World Health Organization. Some of these institutions have also developed curricula that include basic disaster epidemiology, information systems for disasters and communicable diseases, immunization, emergency food programs, and reproductive health.

A refugee crisis in western Sudan (Darfur) [21], a catastrophic tsunami in the Indian Ocean [4], investigations of deliberate biologic and chemical weapons releases, and earthquakes in California [11] may not have much in common, but in the investigation of all four, the epidemiologic approach has proved powerful. As relief agencies have come to accept the role of epidemiology in disaster responses, their reliance on the ad hoc crisis management approach has lessened. An organized approach to data collection in disaster situations also helps disaster managers make crucial decisions and predict the variety of options that they will face during the different phases of a disaster [3].

In summary, with both disasters and the number of people affected by such events on the increase, the importance of disasters as a public health problem is now widely recognized. This is now reflected by the fact that every school of public health in the United States now offers some training opportunities in the public health consequences of disasters (up from just 10 percent 8 years ago). Although all disasters are unique, some similarities exist among the health effects of different disasters, which, if recognized, can ensure that health and emergency medical relief and limited resources are well managed.

REFERENCES

[1] Lechat MF. Updates: the epidemiology of health effects of disasters. Epidemiol Rev 1990;12:192–7.

[2] Logue JN, Melick ME, Hansen H. Research issues and directions in the epidemiology of health effects of disasters. Epidemiol Rev 1981;3:140–62.

[3] Ahern M, Kovats RS, Wilkinson P, et al. Global health impacts of floods: epidemiologic evidence. Epidemiol Rev 2005; 27:36–46.

[4] Shultz JM, Russell J, Espinel Z. Epidemiology of tropical cyclones: the dynamics of disaster, disease, and development. Epidemiol Rev 2005;27:21–35.
[5] Ebi KL, Schmier JK. A stitch in time: improving public health early warning systems for extreme weather events. Epidemiol Rev 2005;27:115–21.
[6] World Health Organization. Rapid health assessment protocols for emergencies. Geneva, Switzerland: World Health Organization, 1999.
[7] Hlady WG, Quenemoen LE, Armenia-Cope RR, et al. Use of a modified cluster sampling method to perform rapid needs assessment after Hurricane Andrew. Ann Emerg Med 1994; 23:719–25.
[8] Malilay J, Flanders WD, Brogan D. A modified clustersampling method for post-disaster rapid assessment of needs. Bull World Health Organ 1995;74:399–405.
[9] Noji EK. Progress in disaster management. Lancet 1994;343: 1239–40.
[10] van den Berg B, Grievink L, Yzermans J, et al. Medically unexplained physical symptoms in the aftermath of disasters. Epidemiol Rev 2005;27:92–106.
[11] Ramirez M, Peek-Asa C. Epidemiology of traumatic injuries from earthquakes. Epidemiol Rev 2005;27:47–55.
[12] Liu S, Quenemoen LE, Malilay J, et al. Assessment of a severe weather warning system and disaster preparedness, Calhoun County, Alabama, 1994. Am J Public Health 1996;86:87–9.
[13] Staes C, Orengo JC, Malilay J, et al. Deaths due to flash floods in Puerto Rico, January 1992: implications for prevention. Int J Epidemiol 1994;23:968–75.
[14] Armenian HK, Noji EK, Oganessian AP. Case control study of injuries due to the earthquake in Soviet Armenia. Bull World Health Organ 1992;70:251–7.
[15] Noji EK. Analysis of medical needs during disasters caused by tropical cyclones: anticipated injury patterns. J Trop Med Hyg 1993;96:370–6.
[16] Burkholder BT, Toole MJ. Evolution of complex disasters. Lancet 1995;346:1012–15.
[17] Boss L, Toole MJ, Yip R. Assessments of mortality, morbidity, and nutritional status in Somalia during the 1991–1992 famine: recommendations for standardization of methods. JAMA 1994;272:371–6.
[18] Public health impact of Rwandan refugee crisis: what happened in Goma, Zaire, in July 1994? Goma Epidemiology Group. Lancet 1995;345:339–44.
[19] Noji EK, Toole MJ. The historical development of public health responses to disaster. Disasters 1997;21: 366–76.
[20] Salama P, Spiegel P, Talley L, et al. Lessons learned from complex emergencies over past decade. Lancet 2004;364: 1801–13.
[21] Guha-Sapir D, van Panhuis WG, Degomme O, et al. Civil conflicts in four African countries: a five-year review of trends in nutrition and mortality. Epidemiol Rev 2005; 27:67–77.
[22] Me´decins Sans Frontie`res. World in crisis: the politics of survival at the end of the 20th century. New York, NY: Routledge, 1997.
[23] Vasterman P, Yzermans CJ, Dirkzwager AJE. The role of the media and media hypes in the aftermath of disasters. Epidemiol Rev 2005;27:107–14.
[24] Galea S, Nandi A, Vlahov D. The epidemiology of posttraumatic stress disorder after disasters. Epidemiol Rev 2005; 27:78–91.
[25] VanRooyen M, Leaning J. After the Tsunami—facing the public health challenges. N Engl J Med 2005;352:435–8.
[26] Hatch M, Ron E, Bouville A, et al. The Chernobyl disaster: cancer following the accident at the Chornobyl nuclear power plant. Epidemiol Rev 2005;27:56–66.

© 2006 IOS Press. All rights reserved.

Constructing a NATO Supercourse

R.Talshinsky, Azerbajan; R. Azmi, Egypt; R. Adlas, U.Kedars, Estonia; L. Bakanidze, Georgia; S.Linn, Israel; A. Rossodivita, Italy; K. Shishani, Jordan; M. Busmane, Latvia; V. Grabauskas, D. Jankauskas, V. Mireckas, R. Obrikis, S. Sliaupa, S. Starkuviene, D. Vaitkaitis, G. Zukauskas, Lithuania; L. Galatchi, Romania; N. Puchkina, E. Shubnikov, Andrey Trufanov, Russia; H. Ghannem, Tunisia; Y.I. Özden, A.E. Onale, Turkey; N. Gudzenko, B. Ledoshchuk, N. Vynograd, Ukraine; J. Dorman, R. LaPorte[a], F. Linkov, E. Noji, J. Powell, P. Rumm, E. Tseytlin, C. Volz, USA

[a]*Department of Epidemiology, Graduate School of Public Health, University of Pittsburgh, 3512 Fifth Ave, Room 310, Pittsburgh PA 15621 USA (e-mail: ronlaporte@aol.com)*

Abstract There is a critical need in NATO countries for better communication in public health and prevention across countries. In addition, there is a very important need to cooperate, communicate, and educate among the different “silos” of information including public health in academia, public health centers, ministries of health, the military and industry. The technologies used to share information now among groups are in many ways not much different than those used 50 years ago. We propose that we need to establish a Prevention Supercourse for NATO countries so that we can seamlessly share prevention information across our “stove piped” prevention information systems. For example, should an earthquake hammer Latvia, it would be essential to obtain the best possible information about earthquakes from the military in France, the Red Cross in Italy, and structural engineers in Estonia. Building a NATO network for disasters and prevention would produce rapid sharing of information for the emergency response as well as educational materials which could reduce the “epidemiology of fear.” In the current paper we outline how this could be constructed.

Keywords. NATO, Internet, Communications technologies, Supercourse

Introduction

The current article overviews the historical development of the Supercourse. As part of this we put out a call for a NATO Supercourse, the goal of which is to improve the sharing of public health information across the silos of prevention.

Cross-country and cross-discipline collaboration in health is critical in this borderless world. AIDs, SARS, acid rain, population growth, global obesity, nuclear warfare, and global warming are all part of international health and wellness. The building of a NATO Supercourse to better integrate research, education and application in NATO countries and across discipline would markedly improve global health.

The Parent Supercourse comprises over 2,200 Microsoft PowerPoint lectures on prevention that is housed in a lecture library on the Internet <www.pitt.edu/~super1/>. It is an open source system where scientists across the world share their best lectures for free and every lecture is "copy left," instead of "copy right," and thus usable by anyone. We will bring the concept of a Supercourse prevention model to NATO Multinational Health, research and education. We will work across NATO countries and disciplines to establish a NATO Supercourse for health. The proposed NATO Multinational Health research and education Supercourse will be used to "Whisk[ing] Research into the Classroom" as described below.

Whisking Research into the Classroom
21 July 2001 BMJ [1]

The marriage between research and teaching has been a loving one. Information from the laboratory is translated to the classroom. It increases wisdom for future generations. Conversely, student questions sharpen our research.

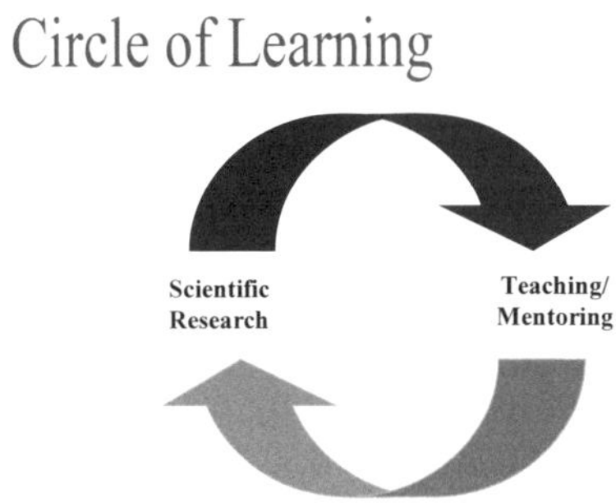

Figure 1: Circle of Learning

Disappointingly, translation of information from research to class meanders. One pathway has research re-written in books and then lectures. This is characteristic of undergraduate training. A second route has faculty creating their own lectures from original articles, as in graduate teaching. Both approaches have worked well. Now they are ill-equipped to handle the speed and quantity of science; Research and teaching are growing distant.

The lack of progress is seen by contrasting research 150 years ago with today. John Snow in 1854 characterized cholera in London. The results were published a year later, and by 1856 the information was in undergraduate curricula—only two years after the original epidemic. In contrast, research completed in 2004 may not be seen in classrooms for more than five years.

The second problem of writing lectures from the original literature is the information explosion. A typical introductory course has 30 topics. It is impossible for faculty to keep up with five areas. Students do not receive state-of-the-art lectures.

A science-to-classroom knowledge speedway needs to evolve. The Internet would be its backbone to speed actionable state-of-the-art research to teachers, and return feedback to scientists.

One such method is a free shareware system called the Supercourse <http://www.pitt.edu/~super1/>. Scientists put lectures in a Web lecture library. This currently has over 15,500 (currently 31,000) faculty from 151 countries contributing 1800 (currently 2550) lectures. Similar models could be developed for any discipline. This model can bridge the "translation gap" between research and teaching.

In order to develop a NATO Supercourse, we first need to recruit experts to the cause. We defined E-recruiting as the methodology associated with the use of the Internet to rapidly recruit faculty and students into the project or activity. In little over five years, we jumped from 100 members of the Supercourse network to almost 29,000 in June 2005. In addition, we have applied the same recruitment approaches in our Department of Epidemiology and we have seen close to a tenfold increase in the numbers of accepted students. Our recruitment approaches are thought of so highly that we were invited to write an article in the prestigious journal *Nature* [2]. We developed new and exciting tools for recruitment and we are planning to utilize them in this project.

Just-in-Time Lecture (JIT)

One means to speed lectures into the classroom is what we call a JIT lecture. As NATO Multinational Health issues hit the newspapers, within days we will put up a cutting-edge lecture on that topic produced by some of the leading scientists.

We have established several JIT disaster lectures. One of our first JIT lectures was developed when, in the week following 9/11, a 14-year old boy crashed his small plane into a bank in Tampa. The lecture was on airline safety and it was designed to reduce the "epidemiology of fear" surrounding the event. Then, we did the same thing with SARS, where a leading SARS expert put up a lecture within a few days after the first case was diagnosed [3]. This lecture was continuously updated and maintained until the SARS epidemic subsided. We published a paper in the *Lancet* on the JIT SARS lecture. We then followed this up with a Bird Flu lecture. After this, was the Bam earthquake in Iran. One of the leading experts on earthquake epidemiology, Eric Noji, provided a general overview of earthquakes in a lecture. Following this Ali Ardaldan, a Ph.D. student at Tehran University, developed a JIT lecture within days after the Bam earthquake. It was a beautiful lecture showing the impact of the earthquake [4]. As a tornado is spotted and confirmed we take our core lecture on tornados and "mass customize it," by adding slides specific to the event at hand. JIT lectures improve the sharing of the best information from the scientists to the classroom. Our latest JIT lecture was dedicated to educating the world about the science of Tsunami, following the events of December 2004. We propose building a NATO JIT program such that if disasters were to occur in NATO countries, almost immediately lectures would be provided in multiple languages to calm the fears of people by providing the population the best possible scholarly information.

Open Source

The final and perhaps most significant aspect of our work is the application of the open source system, developed in the area of software engineering, to the area of curriculum sharing. Open source has led to a very rapid development of software. Our open source approach is not unique. The Human Genome Project, in many ways, is quite similar to our Supercourse of science. The Human Genome Project was the largest and most successful open source project in science, to date. The Human Genome Project funded centers across the world to collect DNA data. These data were almost immediately put on the Internet and shared with the world. The Human Genome Project demonstrated that open source can work at a grand scale. It will also work with a NATO Supercourse.

If we marry NATO Multinational Health research and education to the Internet within NATO countries and make this information accessible, we can have a profound affect on these countries through improved communication of NATO Multinational Health research and education information.

Network Building

Our original parent grant was funded three times by NASA and then through a RO1 by the National Library of Medicine (NLM). From the parent grant, we built our global network of prevention scientists. Here, we focus exclusively on NATO Multinational Health research and education.

The major strength of our research is the network of top scientists that we have already established. What will make this project successful is the building of an international effort for NATO Multinational Health research and education on the Internet. As part of our overall Supercourse, we have already built a large and vibrant group of scientists. In the parent Supercourse we have networked over 28,000 participants from 151 countries. We now have over 2,200 lectures.

Internet and NATO Multinational Health, Research and Education

The speed at which the information revolution is occurring is remarkable. Since 1975, the speed and memory of computers has increased 1 million-fold and prices have plummeted. A $2,000 PC is equivalent to a $10 million supercomputer in 1975. The *Economis*t pointed out that if automobile technology improved as rapidly as information technology (IT), a car would speed along at 100,000 miles per hour, get 200,000 miles per gallon of gas, and cost $5. Acquisition and dissemination of knowledge in NATO Multinational Health research and education has not been as rapid as IT.

Background

Remarkable improvements in NATO Multinational Health research and education have occurred in the U.S. and across the world this century. Most improvements in NATO Multinational Health research and education rely on information sharing. The first two eras of public health in this century were of sanitation and immunization. We argue that the third age will be that of information.

Our philosophy is that the new information technologies have a significant role to play in NATO Multinational Health research and education. We have coined the word "telepreventive medicine" [5]. This discipline uses the tools of the Internet to share information from large numbers of people (both healthy and sick) to prevent disease. It is different from telemedicine, which targets expensive telecommunications technologies to a small number of sick people to "cure" disease. We believe that a "telepreventive medicine" approach will have a much greater effect on improving the health of the world than that of telemedicine. However, the field of "telepreventive medicine" is in its infancy—transnational "telepreventive medicine" barely exists.

Globalization of NATO Multinational Health Research

Increasingly, health concerns are global. Melanomas found in Auckland and California Destruction might be related to the destruction of the ozone layer, global migrations brings new patterns of disease, terrorism, and risk factors such as SARS, world trade of tobacco impacts risk, Despite the globalization of disease, research and education in the area of NATO Multinational Health research barely exists.

The Internet is the most powerful tool for global research, communication, teaching and the translation of research into the classroom. It is a transparent and cost-effective medium, and it is becoming ubiquitous. In addition, the format of the system itself with hypertext links, and "point and click" can lead to powerful new modes of cognitive-based training, as will be described. It is a user-friendly medium and it breaks down the hierarchy between professor and student, and for example, students in Asia, and students in Latin America. On the Internet, people are equal. With the power of the Internet, we can use the best scientists of the world to teach the best talents to foster NATO Multinational Health research and education.

Distribution Systems

It is critical to be able to distribute the global research lectures across the digital divide. We discovered early on that it would not be possible to put the lectures up exclusively in MS PowerPoint format. This was dramatically shown to us when researchers in Ghana tried to pull down one of our lectures and it took 18 hours. In addition, we discovered that the majority of lectures at the Centers for Disease Control and Prevention (CDC) could not be seen in Africa. Instead, we developed a system where we convert the PowerPoint slides into GIF images. The GIF images can be directly used in PowerPoint.

Even with the reduction of file size some countries find it difficult to surf in the U.S. domain. For a Chinese investigator, it costs three times more to visit a site in the U.S. than in China. We solved this problem by building mirrored servers in 45 countries across the world. Most are in developing countries, the last five mirrored servers were put into Mongolia, the Sudan, Malaysia, Romania, and Papua, New Guinea. The site in Pittsburgh gets about 10 million hits a year. In Norway, the mirrored server is hit about 4 million times. We estimate that the approximate number of hits per year is 75 million, making us one of the highest hit sites in health.

We created a CD-ROM containing 1,038 lectures, which also included CDC's Epi Info. We will distribute the CD-ROMs, including the NATO Multinational Health, research and education lectures, to 10,000 participants worldwide. This is a gift that is meant to be given, where people can copy the CD-ROM for at least five others. The CD-ROMs have been distributed to all libraries in the Eastern Mediterranean region of the World Health Organization (WHO). The CD-ROMs are widely distributed across the Former Soviet Union and Pakistan.

World Health Organization (WHO), Pan American Health Organization (PAHO), and Eastern Mediterranean Region (EMRO) of WHO

We have had extensive collaborations with PAHO and WHO. We have worked closely with WHO EMRO regions to distribute the CD-ROMs to every environmental and health library in the region. The Environmental Health Program at WHO Geneva established a WHO Environmental Health Supercourse. The original Supercourse concepts were developed at PAHO, and we have the History of PAHO lecture on our site. In addition, Carlos Gamboa, M.D., a former high level PAHO official, will join us on the task force. We are now discussing with WHO the sharing of our lectures throughout their system. We have over 20 faculty from WHO and PAHO. In addition, we have been in contact with over 1,000 WHO Collaborating Centers, and we have started and taught in over 20 WHO seminars. Our first five papers and many other papers include leading authors from WHO and PAHO. We will use all these approaches to distribute the NATO Multinational Health research and education Supercourse worldwide.

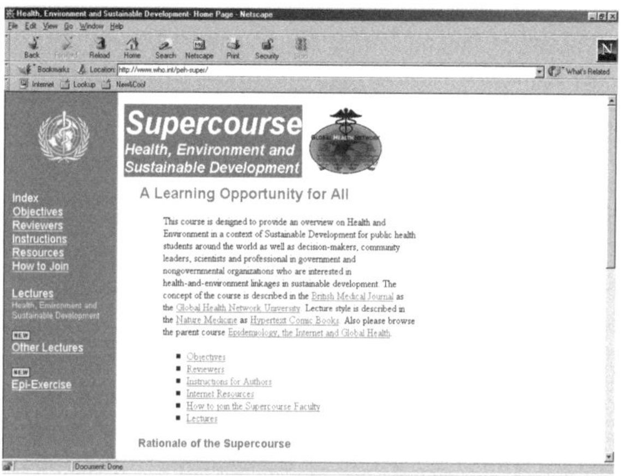

Figure 2: Supercourse website

Progress Report: Global Cooperation

We wrote in *Nature Medicine* about the need for translating and sharing of academic NATO Multinational Health research and education resources. The same concept will be directly applied about sharing NATO Multinational Health research and education resources in the area.

Global cooperation in higher education? the 'supercourse' [6].

A clarion call was just issued by *Nature Medicine* for cooperation in global science. It argued that cooperation prevents duplication and saves precious resources. We suggest that a similar summons should be made for cooperation in biomedical higher education, as this, too, will help all scientists and scholars worldwide. We present here a model demonstrating the feasibility for such cooperation.

Most higher education is traditional classroom instruction. This "information technology" was developed 2,000 years ago. Classroom teaching has survived through the years because it works and it is cost effective. Despite this, classroom instruction is also redundant and inefficient, with considerable waste. For example, a lecture called "Introduction to Epidemiology" has probably been written 5,000 times by 5,000 different people.

How can we eliminate the waste from the existing approach of lecture manufacturing to achieve a lean production system? This can be accomplished with faculty cooperating with other faculty, globally. A new approach, called the "Supercourse," could yield better training for students by improving lectures. Its core is the construction of a free Internet library of lectures. It is based on a model of sharing by software developers called open source, in which software is put on the Internet for all to share and develop. Similarly, lecture manufacturing can be greatly enhanced by having scholars put their best lectures on the Internet for all to see and develop.

It is feasible. The "Supercourse" has in its network more than 1,400 (now 31,000) faculty from 104 (now 151) countries who have contributed more than 100 (now 2,550) lectures and 700 (now 5,000) reviews of the lectures for free. Thus, the "Supercourse" "teaches the teacher" in areas in which they are not familiar, and, therefore, eases lecture development. Moreover, quality is monitored with statistical quality control of Deming. By using low bandwidth, internet scholars in the remotest areas of developing countries can be reached; at the same time, the internet is very powerful, as hypertext links are pathways to different sets

The time of preparation can be reduced by more than 80 percent, and students receive much better instruction. A "shareware library" is a boon to new assistant professors, who start out with no lectures and little time; experienced professors, who have a few dull lectures and; teachers in developing countries who have little access to the biomedical literature. It is time to establish global cooperation among faculty to make our instruction better and less of a burden.

Our philosophy is one of sharing our best PowerPoint lectures with each other. We are a repository of scientific communication, in much the same way as the Library of Congress, or your university library. The lectures are graphically presented research communications, like iconic book chapters [8]. We have been very successful in this area and anticipate equal success in NATO Multinational Health research and education. The latest research can speed from journals into the classroom by direct linkage of lectures with journal articles.

Historical Background

During the initial development of this effort, the Supercourse was funded by NASA three times. One can review the titles of the publications at http://www.pitt.edu/~super1. Our program was quite successful as we have published or submitted for publication 150 articles, 32 of which appeared in *Nature*, *Nature Medicine*, the *British Medical Journal (BMJ)* and *Lancet*. In addition, in 1996, our home page won many awards, the most important being rated in the top 100 home pages by *PC Magazine*, the leading home page in health. The *Lancet* identified our home page as one of the 11 recommended health pages.

We have been fortunate to be collaborating with some of the global leaders in telecommunications, including Tony Villasenor, the former head of the Federal Networking Council of the United States. He helped build our work, and contributed lectures. Vint Cerf, the father of the Internet, has been a strong supporter.

To discuss where we are and where we are going, we first wanted to provide a historical framework for this effort, and then describe the new technologies we have been establishing to demonstrate how this could lead to potentially exciting new applications for NATO Multinational Health research and education.

In 1985, our center became a WHO Collaborating Center. We have published over 300 papers in the area of diabetes epidemiology. At this time, there was a rapid growth in diabetes registries such that centers from 70 different countries joined. This became one of the largest global studies ever completed. Keeping track of the collection of data across the centers was quite difficult. We, therefore, networked as many of the centers as possible smoothing the operation of the project making this WHO project the first to have a telecommunications backbone.

We published the first papers on telecommunications and NATO Multinational Health research in the *American Journal of Epidemiology* and the *Epidemiology Monitor* [7, 8]. We also established the first list-server in epidemiology and public health. We were published as the lead article in the *BMJ* only two weeks after submission [9]. Our Internet articles have been in the prestigious Christmas edition of the *BMJ* four times. One was a lead article [9], and the second was of such interest that the *BMJ* [10] had the editors of the *Lancet*, the *Journal of the American Medical Association (JAMA)*, *Annals of Internal Medicine*, *BMJ*, and others to comment on it. Our article titled the "Death of Biomedical Journals" [11] in the *BMJ* stirred enormous interest and controversy. Tony Delamothe, Web editor of the *BMJ* indicated that the NATO Multinational Health Network was the reason that the *BMJ* moved onto the World Wide Web.

In 1986, we were some of the first people to use E-mail, and then use the Web as a teaching medium. We have a 15-year history of international epidemiology teaching using telecommunication. We published in the *BMJ* a call for the development of a NATO Multinational Health Network University [12].

In 1996, we published a series of review papers in the *American Journal of Epidemiology* [13]. Instead of including tables of reviewed articles, we put these articles onto a Web page, and just included a home page address in the article. Our approach markedly improves the "shelf life" of a scientific review paper [13].

If we use a low estimate of students a year seeing a NATO Multinational Health research and education lecture, in 10 years 200,000 will have seen each lecture. Overall, each lecture would take about $500 to prepare. In 10 years, the cost to present each lecture to a student will be $500/200,000 or 0.25 cents. Contrast this with international "talking head" lectures where costs can run $100/student/lecture or $4,000 for an hour class of 40. Our approach in this case is 400 times less expensive, and highly cost effective, and more accessible to students.

We have the very simple concept that if we can provide excellent template lectures and distribute these NATO Multinational Health research and education lectures to thousands of teachers who can reach millions of students, we can improve the quality of teaching worldwide. We use PowerPoint, as it is the new language of science.

The original Supercourse has been enormously successful for bringing science closer to the training, and for biomedical communication; we see that it will be equally as powerful for NATO Multinational Health research and education. Below we outline some of the features that will be used.

Assist Teachers Worldwide with a Web Lecture Library

Our goal is to provide the best academic template PowerPoint lectures and slides that teachers can use for free to construct their own lectures on NATO Multinational Health research and education. Traditional distance education is not very effective as it doesn't involve the local teacher. In our model, local teachers are central, and their lectures are improved because of our library of lecture templates. Our goal is to assist the local educator not replace them. The majority of the lectures are understandable for those educators having a bachelor's degree. The lectures can be modified by the teachers for K-12 grades. Lectures are easily modified to be personalized for the local context, so that the educators can serve as translators of the topnotch PowerPoint lectures.

Quality Assurance and Statistical Quality Control

Peer Review

Before discussing our work on peer review/quality assurance, it is important to compare what we propose with what is being used in the academia and beoynd. The reviewers might like to examine a recent evidence-based review paper by Tom Jefferson on peer review where he points out that there is virtually no scientific information supporting peer review of journal articles. For lectures on the Web, it is much worse as we know of no lecture series with scientific-based quality control (QC). QC for lectures either does not exist, or it is not scientific. We are just in the process of implementing QC, as the initial stages of our effort has been to create a network, then extract lectures. We present an overview of this, as many scientists become concerned when we are not using the traditional peer review model. Instead, we are employing alternate approaches for monitoring quality that have been highly successful in disciplines such as manufacturing, book manufacturing, and sales. Over the past seven years, a primary area of investigation has been peer review. The peer review issue has been very contentious. Our experience began when we published a paper called the "Death of Biomedical Journals" in the *BMJ* in 1996 [11]. Since then we have been on the cutting edge of the issues related to peer review.

Our approach was originally developed by W. Edwards Deming <http://www.musc.edu/fm_ruralclerkship/cqi.htm> and it has lead to rapid improvements in the quality of most manufacturing goods or services on the market. There is a large corpus of research developing the technology and using this to speed improvement of quality of manufacturing systems. To our knowledge, it has never been used in scientific publishing.

We have five direct and indirect measures of QC ranging from the status of the researcher, the person's university, comments from those looking at the lecture, and a survey at the end of each lecture. The fifth is most important as it is the teachers themselves who receive lectures before presenting. We do not feel that the five prong approach could be considered as a popularity contest. There is a vast difference between our content, compared with reality TV or *Friends*. One reason is that the lectures are being viewed by faculty and students rather than the general public. A second reason is that our Web survey has people rate the content, relevance, presentation, and overall ratings. We also are now asking people about their expectations of the lectures. Unlike other work, we will evaluate the adequacy of the ratings. A Ph.D. student recently completed her dissertation by comparing the surveys with expert ratings by eight senior scientists.

Crossing the Digital Divide

Only 10 percent of the world's people have access to the Internet. We need to reach outside the U.S. to share our science in NATO Multinational Health research and education. We have developed four solutions. The first is to take the picture file (GIF file), and reduce the amount of information using Adobe PhotoShop. A GIF slide is 1/200th the size of a PowerPoint slide. The second is mirrored servers. A mirrored server is a copy of the overview and course located on another computer, but identical to our home page in Pittsburgh. We have 42 mirrored sites including Mongolia, Sudan, Russia, Israel, Czech Republic, two in Germany, Norway, UK, the MOH in China, Hong Kong, Taipei, Japan, Australia, South Africa, Argentina, two in Brazil, Peru, Canada, and the U.S. The third is the distribution of over 10,000 CD-ROMs. The fourth is taking our PowerPoint lectures and making them available as academic comic books.

Lectures and Journals into Developing Countries

Most developing countries have little access to scientific information because of cost. In 1997, in the *BMJ* [15], we were the first to propose the concept of a limited access server, this is a system whereby, say, the *Lancet* could be provided to Mali, Bolivia, or Cambodia for no cost on the Internet, whereas, Japan, the U.S. and the UK would have to pay. The *BMJ* and WHO took up our call, and this year an agreement was signed by six publishers to provide their journals for free to developing countries.

Muslim Collaboration [16]

We have been very concerned as to how to distribute scientific information about NATO Multinational Health research and education in that 90 percent of the world does not have access to the Internet. We, therefore, are examining the possibility of using religions as a conduit for NATO Multinational Health research and education information. All religions have information dissemination systems. In addition, most have, as part of their tenets, hygienic rules. We believe that if we can link the Internet and NATO Multinational Health research and education, we could have the means to bring NATO Multinational Health research and education to many in the last mile.

Current Status and the Origins of the Supercourse

We contacted six Nobel Prize winners as seen below. All six have contributed a lecture, and have been extremely positive concerning the innovativeness of our research. We have found that the Noble Prize winners were quite enthusiastic to join; It is likely that our faculty will be some of the highest powered in science.

We already have a stellar group of scientists. In the past week we started to contact some of the members of the Institute of Medicine, and currently we have 57 who have agreed to contribute lectures.

Figure 3: Nobel Prize Laureates in Supercourse

Our NLM funded Supercourse R01 on prevention has taken off much faster than expected. We would expect similar success in the area of NATO Multinational Health research and education. We have received lectures from some of the world's leading scientists and public health officials, including the Surgeon General, the father of the Internet, and the former head of the CDC. We are likely covering a large percentage of general NATO Multinational Health research and education lectures. From what we have learned, we feel that we can establish a similar type of program in the area of NATO Multinational Health research and education.

With over 150 publications and 32 in upper-tier journals, we have shown that we can build a network, collect superior journals, and publish many papers. The global interest and support is due to the innovative nature of our research. We have a large number of publications in the works.

We have already collected 61 lectures in the area of NATO Multinational Health, a listing of which can be seen at http://www.pitt.edu/~super1/ under NATO Multinational Health Research and Education. The lectures have come from experts at WHO, CDC, universities, developing and developed countries. We have thus laid the foundation for global training in the area of NATO Multinational Health research and education.

Plan of Study for a NATO Multidisciplinary Supercourse

We have a very straightforward plan of study based upon the very successful program of the Supercourse. It is important to spend a few minutes to lay out the model. We will first build an academic network of researchers in NATO Multinational Health using the e-recruiting system that we have devised. Secondly we will extract from the network the best cutting edge lectures on global research. Again it must be pointed out that we developed the systems of e-recruiting, and have some of the best recruiting rates in academia. Moreover we have a very broad dissemination as demonstrated by the parent Supercourse.

Build a Network of Academic Scientists Engaged to Build a NATO Supercourse

The first step is to expand our network of NATO Multinational Health scientists across disciplines. As demonstrated in our report in *Nature*, we have built one of the first very successful E-recruiting approaches. We will now build a repository of NATO Multinational Health research and education research lectures and make these available in an open source library. We have had considerable experience in building the NATO Multinational Health Supercourse network—the 29,000 scientists from 151 participating countries represent close to 80 percent of the countries of the world. Our approach will be different than what we have done before in that we will contact a large number of medical schools in the world to obtain the name of teachers of research design, public health and medical libraries. We have been able to demonstrate a very high rate of participation, as seen in the number of people who joined our network. We will specifically target research methods, and overviews of the existing global research. This team will help us in our goal of recruiting experts into the Supercourse; They will guide us. In addition, they have agreed to help recruit others elsewhere into the Supercourse.

The E-recruitment approach that we highlighted in our *Nature* publication consists of the steps outlined below. These methods have been rigorously tested in several recruitment efforts and proved to be highly effective.

We view our Supercourse as a library. In libraries, old books are not thrown away and are not modified. Modern libraries are not making efforts to update the works of Einstein or modify the plays of Shakespeare. By looking at the date of publication, readers make decisions whether the information presented in the book is acceptable. The Supercourse is similar—although we do our best to keep our lectures updated by networking the authors, we recognize that some lectures may escape the updating mechanisms

Extracting Lectures from the Network

From the network, we will extract academic lectures about NATO Multinational Health research and education. The NATO Multinational Health research and education Web Lecture Library will contain overviews and lectures on 500 topic areas in NATO Multinational Health research and education. The Supercourse is really a set of lectures, not a course. We reduce the burden on the people contributing the lectures. All they need to do is to take the PowerPoint slide, and annotate each slide as to what they would say about that slide. We are quite skilled in identifying lectures. This will be done through Google search, using our network and potentially through our network.

Slides-Lectures-Courses-Curricula

There is the need for higher order curricula building. We plan to take several approaches. One approach will be to use our existing means of organization, including the public health classification, and affiliation classification systems. We will also collect the building blocks of curricula. Curricula comprise courses, courses are composed of lectures, and lectures are composed of slides, as seen in the Figure 4. By re-arranging lectures one makes different courses and curricula.

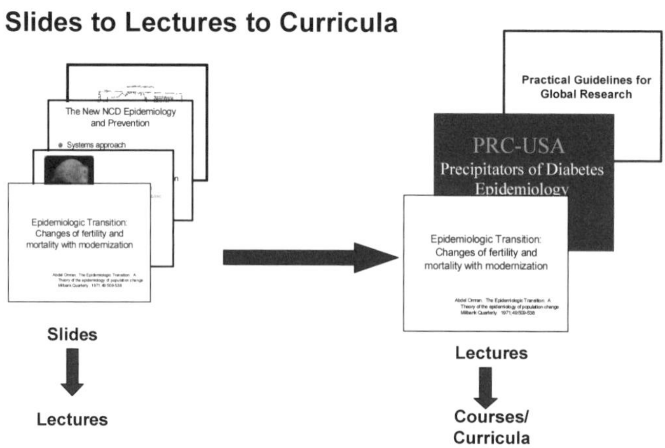

Figure 4: Slides to Lectures to Curricula

We want to improve the education of as many of our future scientists as possible and to help them interact and learn from the diversity of science. We will do this by establishing a system of slides and lectures which can be re-arranged based upon the preferences of local teachers to establish different curricula. This is a very flexible system. Using curricula as the unit of educational material transfer is too static and not sufficiently contextual sensitive. There is no evidence to our knowledge that this can be done on a broad scale. Also, if one teaches a class on behavioral science, the selection of lectures and slides would be quite different from those selected for physics or nursing students. Providing slides and lectures are the “seeds” of interdisciplinary courses and curricula as they make it much easier for local instructors to teach in other areas. We provide the building blocks for lectures, courses, and curricula in the Supercourse. We developed a Google search engine that can search on slide, lecture and course.

It important to recognize several points. First, this system must be better organized than in the pre-Internet days. Also, it must be better organized than the chaos of lectures that currently exists on the Internet. In addition, the Supercourse is targeted towards educators. When we educators give presentations we typically have a general schema already in our heads. We are not blank slates. We will be evaluating all these systems, but it is very likely that we educators will be able to organize lectures as well. The second approach will be to collect complete courses. We already have begun to accomplish this with over 10 complete courses available. Finally, we want to especially help the beginning teacher with "course structures" of excellence. We will therefore begin to collect syllabi and make these available on the Web. So, if a teacher in Zambia needs to teach a course on international research design, he or she might have 10 different syllabi to chose from, as well as the ability to directly contact the experienced teacher.

As we gather more and more syllabi and lectures, these can be shaped into complete curricula. Here, too, we can take two approaches: the first will be to employ our scientific-teaching expert committee; the second will be to gather curricula from masters and Ph.D. programs on the Web.

Lecture translation

Historically, lecture translation in the Supercourse was carried out for free by the Supercourse faculty members worldwide. Instead of centralized and costly translation effort, global faculty members were translating the lectures that were most appropriate for teaching in their region. This approach saved a considerable amount of money and time on the part of the investigators. For example, the Supercourse's Golden Lecture of Prevention that came out in 2003 was translated for free into 14 languages including Spanish, French, Russian, and Chinese.

We recognize that there is a need to help faculty members whose English proficiency may be limited. We are planning to harness the power of machine translation to achieve this purpose. We discussed the power of machine translation in our article in *BMJ* in late 1990s. Machine translation underwent significant improvements since than. Google search engine currently utilizes powerful machine translation software. We are developing similar capabilities for our NATO Multinational Health resource.

Lecture acquisition

There are numerous lectures that could be acquired. The lectures are not designed to replace all of teachers' curricula, instead they are designed to assist and enhance. Thus if next year an educator would like to present on the global burden of disease, the Supercourse would have lectures to use. We recruit topnotch people rather that specific lecture titles. Also, we encourage several authors write on the same topic. Thus, if we have 10 different scientific lectures on global AIDs, one could synthesize these for a presentation.

These NATO template lectures will be directed towards the classroom educators who are teaching beginning level students. The target group will be new public health students, young medical and veterinary students, nurses, pharmacy students, etc., worldwide. However, once established, we will expand to include material for more advanced students and research professionals. Additionally, teachers for K-12 would modify the lectures appropriately for the target population.

Lecture Updating

There is no literature, to our knowledge about lecture updating. This is a description of what we think is a best practice. We will continue to evaluate this to make it better using Continuous Quality Improvement. Based upon five years experience, we have come to a somewhat surprising conclusion that the turnover of information in an epidemiology lecture is small. After five years approximately 90 percent of the material is still viable. A lecture, in many ways, has the turnover of a book in a library, where five-year science books still have considerable material that is usable. Additionally, it is extremely difficult to track down authors to update lectures.

We have developed several strategies to increase the shelf life of lectures. The first is redundancy. We encourage different lectures on the same topic. Thus, Sir Richard Doll may deliver a lecture from 1995 on smoking and health, and in 2001, Lucile Adams would produce a lecture on smoking and health. The user can pick and choose slides from either presentation to build a "fresh" lecture. A second approach that we plan to build is for each lecture, and each slide, to have a "pocket," where one can put comments or new slides. Thus, for Doll's lecture on smoking and health one could search to find any comments or figures by full professors of epidemiology that were posted after January 1, 2002. This has some of the characteristics of Amazon.com and Wikipedia, a free encyclopedia that can be updated by users.

The best content on NATO Multinational Health research and education in the world is of little use unless it is distributed. We will, of course, use our Web page http://www.pitt.edu/~super1 to distribute the Supercourse. In addition, we have 42 mirrored servers worldwide for people to access the lectures within their own countries. We recognize, however, that the Internet covers only about 10 percent of the world. We, thus, established a system of CD-ROM distribution for the course. We plan to include the NATO Multinational Health research and education lectures as a part of this. Last year, we distributed about 10,000 CD-ROMs to academic scientists worldwide, and asked them to distribute the CD-ROMs to at least five people they know. Our procedure is: to first drop our Supercourse with the expanded lectures into servers in the other countries, and every three months send new files to the mirrored servers to update the mirrored servers with new lectures.

The approach that we will be taking to distribute lectures will be somewhat different than what we are doing now. We will first identify the entire medical (and potentially nursing, dental and veterinary schools) in the world. We will locate and contact people who teach research design, and will share our lectures with them, and then follow up to see how many people have used our lectures.

References

[1] Ronald LaPorte, Akira Sekikawa, EunRyoung Sa, Faina Linkov, and Mita Lovalekar Infopoints: Whisking research into the classroom BMJ, Jan 2002; 324: 99.

[2] LaPorte, R Recruiters and Academia Nature 428,876 (2004)

[3] Chotani RA, LaPorte RE Linkov F, Dodani S, Ahmed D, Ibrahim KM. Just-in-time lectures: SARS. Lancet 2003 361:1996.

[4] Ardalan A, Linkov F, Holakouie K, Noji E, LaPorte R Bridging Schools of Public Health between Iran and the US: Teaching and sharing about the Bam Earthquake. Lancet 2004:9423

[5] LaPorte RE for the NATO Multinational Health Network. Commentary: Telepreventive Medicine-the Autobahn to Health. Brit Med J 1996;313:1383-1384.

[6] Supercourse Faculty. Global Cooperation in Higher Education. Nature Medicine, 2000, 6, 358

[7] LaPorte RE for the Global Health Network. Commentary: Telepreventive medicine - the autobahn to health. Brit Med J 1996;313:1382-1384.

[8] LaPorte RE, Gower IF, Rewers M, Tuomilehto J, Tajima N, Akimoto Y, Aiso M, Grabauskas VJ, Williams JG. Telecommunication and International Health Research. Am J Epid 1988;128(2):439-42
[9] LaPorte RE, Tuomilehto J, Tajima N, Dorman J. Multinational Networking of Epidemiologists {Letter}. The Epidemiology Monitor 1986;7(10):2-3.
[10] LaPorte RE, Akazawa S, Hellmonds P, Boostrom E, Gamboa C, Gooch T, Hussain F, Libman I, Marler E, Roko K, Sauer F, Tajima N. global Public Health and the Information Superhighway. Brit Med J 1994;308:165.
[11] LaPorte RE, Akazawa S, Gamboa C, Maclure M, Marler E, Sauer F, Shenton C. The Death of Biomedical Journals? Brit Med J 1995;310:1387-90.
[12] LaPorte RE, Akazawa S, Boostrom E, Campos M, Gamboa C, Gooch T, Lee HK, Libman I, Marler E, Roko K, Sauer F, Tajima N, Wiebe W. global Public Health and the Information Superhighway: NATO Multinational Health Network University Proposed. Brit Med J 1994;309:736-7.
[13] Chang Y-F, Portuese E, LaPorte RE. Increasing the shelf-life of review papers {Letter}. Brit Med J 1996;312:513.
[14] Linkov R, LaPorte RE. Can ebooks help to cross the digital divide? (letter) BMJ Oct. 18, 2002 LaPorte RE for the NATO Multinational Health Network. Internet server with targeted access would cure information deficiency in developing countries. BMJ 1997;314:980.
[15] LaPorte RE for the NATO Multinational Health Network. Internet server with targeted access would cure information deficiency in developing countries. BMJ 1997;314:980
[16] Husseini A, LaPorte RE. Islam with the internet could do much to prevent disease. BMJ 2001; 323:694.

© 2006 IOS Press. All rights reserved.

Former Soviet Union (FSU) Supercourse Developments

Eugene SHUBNIKOV, MD[1]
FSU Internet Prevention Group

Abstract. This paper explains the roots and aims of the Internet Prevention (I-Prevention) program created in the frame of Supercourse project—online library of lectures in public health, epidemiology, and global health <http://www.pitt.edu/~super1>. We present a new, multinational project in public health for former Soviet Union (FSU) countries, which should help these newly independent countries create their countries' public health and prevention programs on the basis of the best knowledge available. The global participation through an Internet-based disease prevention system is the primary goal of the research. We networked scientists involved in prevention and the Internet in Russia and FSU countries. We established a bilingual set of lectures on the Internet on prevention medicine. We created a Russian language library of lectures as a backbone for a Russian language "telepreventive medicine" program.

Keywords. Public Health, Russia, Russian Internet, Supercourse, life expectancy

Introduction

In most developed and developing countries there has been a 25-year increase in life expectancy during the last century. This increase is greater than that seen at any time in history. The increased longevity is almost exclusively due to public health. Public health professionals and workers have many reasons to be proud of their accomplishments. It has been estimated that almost 24 of the 25-year increase was due to simple public health measures such as sanitation, immunization, changes in lifestyles, improved nutrition, etc [1]. Unfortunately, morbidity and mortality from cardiovascular and other non-communicable diseases are high in Russia with no evidence that they are declining [2]. Much of this is due to high-level risk factors among the Russian population including unbalanced (unhealthy) nutrition, alcohol consumption, and smoking. Additionally, increased drug abuse and HIV infections play a determining role. In early 1999, 2 percent of Russia's population was officially registered alcoholics [3]. Most certainly, this is an undercount. Russians now consume more alcohol per person than probably any other country in the world. The average Russian male possibly drinks twice of what an average American male drinks. Alcohol consumption is a major factor in lowering male life expectancy to a very young 57 years [4].

[1] Corresponding author: Eugene Shubnikov, MD, Institute of Internal Medicine, Research Scientist, P.O.Box 668, Novosibirsk, 630090, Russia. eshub@mail.nsk.ru.

Longevity today differs little from life expectancy right after the Second World War. Additionally, smoking is very common among both males and females. Among males ages 25-34 years, 73 percent smoke and 41 percent smoke who are ages 65 or older. Among females, smoking is much more common among the young, ages 18 – 34, of which 27 percent smoke, than among middle-aged and elderly, ages 55 years and older women, where five percent smoke [5].

Drug addiction and the exchange of contaminated needles, associated with drug abuse, have resulted in one of the world's fastest-growing rates of HIV infection [6]. Some medical devices as gastro scopes, dental care procedures, and even using non-sterilized equipment at barbershops has also contributed to the growth of HIV infections in Russia. Thus, the reduction of life expectancy in 90s can be attributed to changing economical, living and working conditions, and unhealthy life styles.

Many years of neglect of medical services during the Soviet regime led to the current state of ineffective and outdated facilities and practice. Further changes in the economy have lead, in part, to a breakdown of public health. Clearly, there are problems, but there are solutions as well.

The Problem: Health Status Background of Russian Federation

It is useful to begin with an account of health status developments with a consideration of the Soviet period, as the present health crisis of the Russian Federation has its roots in events that long precede the collapse of the Soviet Union [7].

At the end of the 19th century, life expectancy in Russia was 32 years, compared to 47 years in France and the United States. By 1938, the gap had increased: life expectancy was 43 years in Russia, while in France it had reached 59 years, and in the United States over 63 years. In the course of roughly the next 15 years, however, Russia made huge strides and succeeded in closing much of the gap. By 1965, life expectancy for males was 64.3, 67.5, and 66.8 for Russia, France, and the United States, respectively, and 73.4, 74.7, and 73.7 for females [8].

Following the establishment of the "Semashko" model in 1918, Russia made massive strides in preventing the spread of infectious diseases. Drastic epidemic control measures were implemented particularly in the cases of tuberculosis, typhoid fever, typhus, malaria and cholera. These involved community prevention approaches, routine check-ups, improvements in urban sanitation and hygiene, and quarantines, etc [7].

Progress in controlling infectious diseases continued to be made throughout the entire Soviet period, with crude mortality rates from infectious diseases falling from 87 per 100,000 in 1960 to 21 in 1980, and 12 in 1991 (representing a fall of 86 percent) [9].

Yet, these successes were not reflected in improvements in the overall health status of the population. Whereas, life expectancy in Soviet Russia (and the entire Soviet Union) showed improvement until the 1960s, it subsequently stabilized presaging a growing gap with Western nations that were to experience continuously increasing life expectancies. By about 1990–91, male life expectancy was roughly at the same levels or lower than in the 1960s (63.8 years in 1990 compared to 63.0 years in 1959, and 64.3 in 1965), while female life expectancy had improved only slightly over the 30-year period (71.5 years in 1959, 73.4 years in 1965, and 74.3 in 1990). The variation in death rates after 1965 was particularly unfavorable in Russia's male population [10]. By the beginning of the 1980s the gender gap in life expectancy was about 12 years, which is the largest gap in every country, worldwide [11].

The diverging paths of Russia and other industrialized nations with respect to health status from the 1960s onward has been attributed to the failure of the Russian health care system to successfully respond to the epidemiological transition. In a sense, the very factors accounting for the successes of the early years in curbing the spread of infectious diseases were responsible for the inability to effectively cope with non-communicable diseases that were becoming increasingly important. The Soviet health care system was based heavily on prevention, consisting of extensive screening measures and check-ups, although with little evidence of their effectiveness, which together with the introduction of antibiotics after the Second World War, contributed to major reductions in infectious diseases. This system of prevention, with its primarily medical orientation, did not evolve into one of the population-based health promotion measures necessary for dealing with the new patterns of morbidity (due to non-communicable diseases). While those in power were aware that an epidemiological transition was occurring, they responded by treating cardiovascular and other non-communicable diseases as "social diseases" requiring a medical solution [9].

By the 1980s, the gap between Russia and Western countries in life expectancy at birth came to about 10 years for men and 6 years for women, mostly due to high death rates among those of working age [8].

Thus, towards the end of the Soviet period, the Russian population, faced with mortality rates much higher and life expectancy lower than that of its Western neighbors, was already confronting a health crisis that had been developing gradually over the previous two to three decades [7].

The health status of the Russian population declined precipitously following the collapse of the Soviet Union in late 1991. By all accounts, in the last decade Russia has been experiencing a shock, unprecedented in peacetime, to its health and demographic profiles. During the 1990s, mortality rose for all age groups except infants and children, and for both sexes, though the increase was far greater for males. The greatest increase in mortality was among males ages 40–49, which nearly doubled (87 percent increase) in the period 1990–1994 [12].

The underlying causes of the developments described above have been debated at length, and can be briefly summarized as the impact of a major social and economic shock in a population already vulnerable because of poor diet, high levels of smoking, and weak systems of social support, in which alcohol and, increasingly, intravenous drugs, are easily available. This is in a country where the health care system is poorly equipped to respond to these challenges [7].

The Solution: Improving Health in Russia through Technology

Clearly, the high mortality rate in Russia is avoidable, as large number of deaths in Russia result from a series of preventable diseases. Russia needs an inexpensive but effective approach to prevention, which will improve its system of public health. Currently, traditional sources of public health information in Russia are inadequate; journals and booklets published by special medical departments for prevention are hard to come by because of cost; popular journals, as well as medical professional publications, are filled mostly with advertisements and rarely provide public health educational materials.

Prevention, especially of non-communicable diseases, has not been a major focus of the public health services in Russia. In today's world preventing diseases before they start is much cheaper and easier than dealing with them when it is too late. The Internet, coupled with prevention, promises to be the answer as most of public health involves the transfer of information.

The Internet, a global network of networks, that enables computers of every kind to communicate directly and transparently, and share services throughout much of the world, is enormously valuable. It constitutes a shared global resourcece of knowledge, data, information, and provides means of cooperation and collaboration among the mixed communities of Russia and the world. Therefore, the Internet promises to be an answer to the problems of public health in Russia, as Russia is very well suited for the use of the Internet for improving health.

However, the problem is that the technologies that form the backbone of public health information sharing, even in advanced countries such as Russia, the U.S., and Japan (the telephone, postal systems and fax) become quickly outmoded. Currently, we are in an information revolution where rapid, almost free, sharing of data halfway around the world is available. An enormous advantage for Russia is that the country is one of the most advanced in the world in terms of being wired. Essentially, all areas of Russia have access to the Internet, including wireless satellite Internet. If epidemiologists and public health scientists in Russia, the U.S., and the world were to systematically harness this Internet revolution, which has rapidly taken place in government and private sectors, there could be a rapid expansion of research efforts, a reduction of costs, and an enormously improved translation of findings in public health for current and future generations. The *Economist* magazine stated that with the Internet there is a "death to distance." Thus, we can bring scientists in Russia closer to each other and to other scientists worldwide.

I-Prevention: The Revolution of Public Health in Russia

The initials stages of public health during this century addressed sanitation and immunization. The evolving stage is the exchange of information—information for non-communicable diseases is what vaccinations were to infectious diseases.

To advance online information sharing we have created a new discipline called "I-Prevention" that uses the tools of the Internet to collect information from large numbers of people (both healthy and sick) to prevent disease and disability. Information has always improved health. The Internet provides an inexpensive way for individuals with the common goal of preventing disease and disability to distribute information person-to-person within and among countries. It is different from "telemedicine," which targets expensive telecommunications technologies to a small number of sick people to "cure" the disease. We believe that "I-Prevention" together with "telemedicine" will revolutionize public health in Russia.

Prevention has always been more effective than cure, and should always come first. But people will always get sick so we need "telemedicine" as well. In Russia prevention of non-communicable diseases is mostly dependent on the coordination of the medical community and public health officials from different cities across the broad and large country. Despite the need for coordination, there is little communication throughout the country. Much of Russia is already connected to the Internet; however, the people engaged in prevention have not used it to as such to a great extent.

Development of the Russian Internet

The structure of the Russian Internet began to develop in 1991-1992. As a first step the Internet companies (Eunet/Relcom, Sovam Teleport, Glasnet, FreeNet) only provided E-mail access. In 1993–1994, creation and reinforcement of the scientific and educational networking was established (Radio-MSU, RUNNet). From this period, all Internet companies started to provide access to growing World Wide Web resources. In 1996, the networks Demos/Internet and RELARN-IP were established. Much more attention to the Russian Internet came from the Global One Russia and RosNet companies. In addition RSSI—the Russian Space Agency Internet was established, which has a connection to NASA's Internet.

At the end of 1995, the maximum exchange between Russian network and foreign providers was about 2MBs, but by the end of 1996 it reached 15 MB/c and is continuously growing. According to data of Comcon <http://www.comcon-2.com>, the end of June 1998 there were about 840,000 people who identified themselves as Internet users. Also, there were about 15,000 independent sites on Russian Internet. The Russian Internet Survey <http://www.lexa.ru> revealed that there were already 33,000 independent resources (sites) in September 1999. By the end of 1999, the number of Internet users in Russia reached 2 million people. According to the Dataguest company forecast the number of Internet users was expected to double by the beginning of 2003 (4–4.5 million of users) [13]. As of early 2006, Russia has up to 18 million Internet users.

Regarding science and health, the Soros Foundation <http://www.soros.org> has created a non-profit network for science and education in Russia, and has established computer classes and free Internet access in Russian universities, institutions, and schools. The Soros Foundation supports 33 well-equipped Internet centers at the leading Russian universities. Thousands of hospitals, medical schools, and libraries have been wired with funds from the foundation.

I-Prevention and Public Health Training Program

Application of the Internet to public health in Russia could markedly accelerate the pace of progress in health by bringing epidemiologists and other public health scientists closer together, and to other areas of science and public health. The groundwork has already begun with advanced networking and an educated population. The primary area of application is training. New systems need to be established to train people in the area of prevention and I-Prevention. Recently, we have argued that the best approach for improving training is to improve the "content" e.g., lectures. The Supercourse <http://www.pitt.edu/~super1> has been established as a global system of shareware lectures. This effort was originally funded by NASA, and is now funded by the National Library of Medicine. In 2000, approximately 1,504 faculty from 104 countries have joined, but only five were from Russia. We decided to change this situation.

First, we networked scientists involved in prevention and the Internet in Russia. We used the global model of the Supercourse for recruiting experts who are already connected to the Internet in Russia. They helped to provide lectures, to recruit new I-Preventionists and to use the lectures for training students. Second, we established a large Russian Language Lecture Library on the Internet. This created a backbone for a

Russian I-Prevention program based upon the model established by the Supercourse. This is the first Internet program for prevention of non-communicable diseases among the populations of Russia. Third, we made it possible for Russian scientists to report latest findings for immediate translation into prevention on the Internet.

Therefore, researchers in Russia and throughout the world were able to find lectures and data on emerging and viral diseases, infant mortality, chronic heart disease, cancer, etc., in Russia by region/site. Previously, scientific data was scattered in many different places over the Web with no verification as to the quality of the information.

Implementing Results

We developed a special Russian library of shared Web-based I-Prevention lectures. We now have rich public health information resources that can be used to train large numbers of students in public health topics in Russia by helping the public health teachers. Instructors teaching public health have cutting-edge, interesting lectures available from around the world to help teach public health research and prevention. The creation of new courses in public health for the Internet at Russian medical colleges can be easily developed with the help of the library of lectures, as faculty can use template lectures and add slides.

All Russian university and medical college faculty members who teach components of public health will be given free access to the Web library to teach students. From this library, they can share knowledge, education, and training systems with other public health professionals in Russia and worldwide. Technically, it is already possible as all Russian universities have direct access to the Internet through the satellites and also some commercial companies provide relatively inexpensive access to the Internet from home computers.

Increasingly, health concerns are global. For example, an epidemic of the flu originating from chickens in China can reach Russia, Ghana, Peru, and Japan within hours. HIV infection has been likely spread by infected people migrating to other areas in airlines, drug trafficking in cocaine, and health are international problems. With these issues in mind, it is important to reflect upon the current state of training in epidemiology/global health.

Until recently, there was no formal training in this field in Russia compared to the U.S., where students in schools of public health received and continue to receive excellent training. Now, the St. Petersburg's Medical Academy of Post-Graduated Studies and the I.M. Sechenova Medical Academy of Moscow are well on their way to becoming schools of public health with the help of the Soros Foundation. Russian medical colleges/schools, dental schools, and nursing schools epidemiologic training is limited to infectious diseases [9], but it is cost prohibitive ($30,000 per year) to send Russian students to the U.S., where income per capita is 10 times more. Fortunately, the Soros Foundation has already funded a Russian language textbook on public health titled *The New Public Health*. The Soros foundation supports five new departments in medical schools that will teach public health courses to students and to doctors in Russia, plus two programs that will eventually be full scale public health programs.

We intend to target young medical, dental, veterinary, nursing, pharmacy students, and of course, public health students in Russia. As this system is designed to "aid the teacher of epidemiology/global health"—"teach-the-teacher"—there is the potential of reaching 30–50,000 students each year.

Connecting Russian Scientists to the Global Network

University instructors will select and modify lectures from the Supercourse. This initial material will be directed towards the classrooms of epidemiology/global health instructors who teach beginning level students in various areas of health, biology, and statistics, who have had no training in epidemiology/global health, and know little about the Internet. We will expand to include material for more advanced students and research professionals. Our goal is to make this course sustainable after the lectures are put on the Web. We envision several approaches leading to sustainability, and will allow 24/7 free access to the lectures on the Web. Note that this is not a degree granting course, but we are examining the possibility that lectures could be used in certificate training. The course could also evolve into continuing education.

More Russian Supercourse lectures will be produced by leading scientists in the field of cardiovascular diseases, cancer epidemiology, dangerous infectious diseases, leading nutritionists, and by main scientists from the research institutions that are also recognized as leaders from the world faculty. We are planning to have national and local public health professionals provide lectures such as from the Ministry of Health. The topic of the lectures will include actual health problems in Russia including decreasing life expectancy, incidence of tuberculosis, HIV infections, diabetes epidemiology, and infant mortality. We will have a set of leading specialists in each field who will check the Russian Library of Lectures and who will be a mentor for particular topics. We will have special lectures that will describe health conditions in different parts of Russia to assist regional specialists to in improving public health in their regions. These data will also be very beneficial for the outside world, as it is not clear where to obtain accurate data on health in Russia.

Conclusions

The improvement of public health in Russia requires cost-effective solutions that can be integrated with existing systems. In Russia, many, if not most public health experts have access to the Internet, and can, therefore, use the Supercourse as a resource. However, because there has not been an effort to create a field of I-Prevention, we created the Russian I-Prevention program <www.pitt.edu/~super1/national/index.htm>.

We expanded our network to FSU countries, and now have participants in every FSU country, who are in constant contact with each other through our newsletters and personal contacts. Currently, we have 374 participants from all FSU countries; the mailing list includes more than 450 participants.

So, together with the parent Supercourse project and our Russian Internet prevention program we have created a network prevention program that is now accessible for Russia, FSU countries, and global audiences. The FSU Supercourse has more than 285 lectures by FSU authors or lectures about FSU health status. About half of 285 lectures are available in Russian language. There are also 15 health profiles for each FSU country.

We created several other Web pages including courses and dictionaries, and plan to continue the development of the Web component of our program. The I-Prevention materials may be used in all medical projects in Russia and the FSU. The program will help determine rational spending of limited resources for medicine in FSU countries, and could become the backbone for other health projects, for example, cooperation in the field of prevention of disasters.

References

[1] World Development Report. 1993. Investing in health. Oxford University Press.
[2] Glazunov IS. 1998. Necessity of elaboration of policies and strategies of cardiovascular and other non-communicable diseases prevention. Diseases prevention and health promotion. N1. Available from: www.mediaspera.aha.ru
[3] Available from: http://clinton.cnn.com/WORLD/9601/russia_drunk/ accessed June 1, 2005
[4] Alcoholism imperils Russia Available from: http://www.beachbrowser.com/Archives/news-and-reviews-April-99.htm accessed June 1, 2005
[5] McKee M, Bobak M, Rose R, Shkolnikov V, Chenet L, Leon D. 1997 Spring. Patterns of smoking in Russia. Tob Control 7(1): 22-6.
[6] Available from: http://www.ippf.org/newsinfo/archive/0005/68.htm
[7] Tragakes E. and Lessof S. Tragakes E. ed. 2003. Health care systems in transition: Russian Federation. Copenhagen, European Observatory on Health Systems and Policies. 5(3).
[8] Shkolnikov V., Mesle F, Vallin J. 1996. Health crisis in Russia: Recent trends in life expectancy and causes of death from 1970 to 1993 (Parts I and II) Population: An English Selection. Vol. 8.
[9] Tulchinsky, T.H. and Varavikova, E.A. 1996 March. Addressing the epidemiologic transition in the former Soviet Union: Strategies for health system and public health reform in Russia. Am Journal of Pub Health 86(3).
[10] McKee M. and Shkolnikov VM. 2001. Understanding the toll of premature death among men in eastern Europe. BMJ 323:1051-5.
[11] Shkolnikov VM. and Chervyakov VV. 2000. Policies for the control of the transition's mortality crisis in Russia, UNDP, Russia Transitional Family Research Institute, Moscow.
[12] The United Nations in the Russian Federation, Common Country Assessment, Draft No.1. 2001 March.
[13] Skabichevskiy L. 1998 11 November. "Russia under Network," Kommersant. 208:8.
[14] Vlassov VV. 2000. Is there epidemiology in Russia? J Epidemiology Comm Health 54:740-4.

Scientific Networking and the Global Health Network Supercourse
F. Linkov and R. LaPorte (Eds.)
IOS Press, 2006
© 2006 IOS Press. All rights reserved.

The Use of Just-in-Time Knowledge at the Time of Disaster

Faina LINKOV, PhD[a], Ronald LAPORTE, PhD[a]
[a]*Graduate School of Public Health, University of Pittsburgh, Pittsburgh, Pennsylvania*

Abstract. Hurricane Katrina became one of the most destructive and costliest natural disasters in the history of the United States. In addition to causing death and destruction, Hurricane Katrina, followed by Hurricanes Rita and Wilma, heightened societal fears and misperceptions about the nature of hurricanes. Teachers and public health educators across the country and worldwide had only limited materials to educate their students on the risk and risk factors of hurricanes and disasters. When materials were available they were difficult to find by people outside of the scientific community, and their format was not user-friendly. Overall, the scientific community was poorly prepared to deliver basic scientific facts about hurricanes to their own students, students of the world, and the general public. The Global Network Supercourse group decided to change this situation and deliver a basic lecture on the science of hurricanes to the audience of public health professionals and educators. To address the problem of the "epidemiology of fear" associated with disasters; our group developed several scientific lectures dealing with basic facts about hurricanes, and made them available through the Supercourse network of over 30,000 scientists 151 countries. As scientists and as public health educators, the Supercourse group will continue to deliver cost-effective public health knowledge about the science of natural and man-made disasters to prevent the "epidemiology of fear" from infecting our collective consciousness. What was done by the Supercourse group with JIT knowledge on hurricanes could potentially serve as a model for teaching the public about all natural and man-made disasters through "lectures on demand."

Keywords. Public Health Education, hurricane, JIT knowledge

Hurricane Katrina and the "Epidemiology of Fear"

Hurricane Katrina was the first Category 5 hurricane of the 2005 Atlantic hurricane season. It was the third most powerful storm of the season behind Hurricane Wilma and Hurricane Rita, and the sixth-strongest storm ever recorded in the Atlantic basin. Katrina caused the collapse of several sections of the levee system in New Orleans. It was one of the most destructive and costliest natural disasters in the history of the United States with massive displacement of a U.S. population.

In addition to causing death and destruction, Hurricane Katrina, followed closely by Hurricanes Rita and Wilma, heightened societal fears and misperceptions about the nature of hurricanes. Media reports, pictures of decomposed corpses on TV screens, and rumors increased the "epidemiology of fear" [1]. Despite the fact that the odds of dying in a hurricane remain very low compared to that of motor vehicle accidents, fires, accidental poisoning, or falling [2], the fear of hurricanes replaced rational thinking and scientific facts.

Prior to Hurricane Katrina, the average number of people killed by tornadoes in the past decade was more than twice the number of hurricane deaths, 57 a year vs. 21, according to the National Weather Service [3]. Teachers and public health educators across the country and the world had only limited materials in which to educate their students on the risk and risk factors of hurricanes and disasters. When materials were available, they were difficult to find for those outside of the scientific community, and the format of information was not user-friendly. Overall, the scientific community was poorly prepared to deliver basic scientific facts about hurricanes to students, the students of the world, and the general public.

Model of Supercourse Rapid Development of Hurricane Lectures

The Global Network Supercourse group decided to change this situation and deliver a basic lecture on the science of hurricanes to the audience of public health professionals and educators. The Supercourse is a library of over 2,300 free lectures on prevention, shared by over 30,000 public health professionals from 151 countries [4]. Targeting the educator, the Supercourse aims to deliver top quality lectures to scientists and educators from around the globe. The open access Supercourse Library of Lectures consists of low-bandwidth lectures developed by academic prevention experts in the fields of epidemiology and public health.

To address the problem of the "epidemiology of fear" associated with disasters, our group, in collaboration with Dr. Ali Ardalan from Teheran, Iran, developed several scientific lectures dealing with basic facts about hurricanes. The work of Dr. Ardalan, having the support and input of 30 leading academic meteorologists around the world, encouraged scientific solidarity around the world. This solidarity was leveraged through the network of the Supercourse. We unveiled our hurricane lecture <http://www.pitt.edu/~super1/lecture/lec20371/index.htm> in just several days following Hurricane Katrina. The lecture was quickly translated into Spanish by Dr. Nicolas Padilla, Celaya, Mexico, and Chinese by Dr. Yuan Ying, Peking, China. This lecture was followed by the Hurricane Rita lecture http://www.pitt.edu/~super1/lecture/lec21031/index.htm, and received more than 4,200 hits from around the globe as of November 15, 2005. In addition to JIT lecture on hurricanes, Steven Businger, PhD, professor of meteorology from Hawaii, immediately provided us with his series of hurricane lectures <http://www.pitt.edu/~super1/lecture/lec20671/index.htm>.

Global Dissemination of Hurricane Lectures

Besides developing the educational materials, we needed to find the means to deliver the lecture to the educators worldwide. We sent a note to our network of 8,000 academic faculty in the United States requesting assistance in delivering the lecture to schools throughout Texas. Within a few hours we were contacted by educators who wanted to distribute it. Dr. Steve Warin, University of Texas, was able to E-mail the lecture to every school and university in Texas. In addition, Texas Epidemiologist Thomas Betz provided the lecture to all epidemiologists in Texas.

One day later, Diane Durno from the Allegheny Intermediate Unit in Pittsburgh, Pa., contacted us. Durno distributed the lecture to all of the teachers in Allegheny County, Pennsylvania, and then, throughout the entire Commonwealth of Pennsylvania, distributing the information to a total of 42 states. This all occurred during the prodromal phase of Hurricane Rita.

Within the six days after Hurricane Rita appeared on the radar screen in the Gulf of Mexico, the Global Health Network created two scholarly lectures that were distributed to schools in over 85 percent of the states, with distribution completed two days before Rita hit land. This was done at very little expense by tapping into the network of networks worldwide.

JIT Knowledge to Combat Fear

What is the next step of public health education following this year's hurricane season? Although television screens can scare us with disturbing images of human waste, toxins, and physical violence, infectious diseases are not the major threat of post-hurricane recovery. There is a need to create scientific awareness about the long-term and distant impacts of these disasters, so we can help the community consider and embrace more systematic and long-term solutions for the prevention of the "epidemiology of fear"—although the short term transition may be perceived as more painful because it is more complicated than the "simple" devastation pictured on TV and other media.

JIT knowledge is a tool that can be used by public health educators in disaster areas to prioritize and deliver essential, targeted public health information to those who need it most. Instead of conflicting public health information delivered by the likes of media personality Wolf Blitzer, these lectures are the best science that can be produced. Certainly there will be "uncertainty," which is the nature of science, but teachers and students are obtaining education from reputable sources. Additionally, JIT knowledge has great potential to reduce people's fear of hurricanes and other disasters among those who live in areas that are not prone to be hit by hurricanes. Effective delivery mechanisms for basic disaster preparedness and disaster mitigation information are of key importance for any natural or man-made disaster.

Conclusion

Networks of public health professionals, networks of interdisciplinary professionals, as well as networks of networks, are very powerful means to disseminate disaster prevention and mitigation messages. With the distribution of hurricane lectures, the Supercourse group demonstrated that JIT knowledge can be distributed rapidly and virtually free-of-charge throughout the trusted network of scientists, administrators, educators, and public health practitioners. As scientists and as public health educators, the Supercourse group will continue to deliver cost-effective public health knowledge about the science of natural and man-made disasters to prevent the "epidemiology of fear" from infecting our collective consciousness. What was done by the Supercourse group with JIT knowledge on hurricanes could potentially serve as a model for teaching the public about all natural and man-made disasters through "lectures on demand." The hurricane lectures and other JIT lectures are available on the Supercourse Web site < http://www.pitt.edu/~super1/JIT/jit.htm>.

References

[1] LaPorte RE, Ronan A, Sauer F, Saad R, Shubnikov E. 2002 June. Bioterrorism and the "epidemiology of fear." Lancet Infect Dis 2(6):326.

[2] National Safety Council.What are the odds of dying? [Citied November 3, 2005]. Available from: http://www.nsc.org/lrs/statinfo/odds.htm

[3] Hill M. Few places in United States safe from natural disasters Available from: http://www.azcentral.com/news/articles/1105safeplaces05.html

[4] Global Health Network Supercourse Web site [Accessed October 30, 2005.] Available from: http://www.pitt.edu/~super1

Scientific Networking and the Global Health Network Supercourse
F. Linkov and R. LaPorte (Eds.)
IOS Press, 2006
© 2006 IOS Press. All rights reserved.

The Supercourse Provides Model for a Global Safety System Prevention Network

Ronald E. LAPORTE [a], Andrey TRUFANOV [b]
[a] *University of Pittsburgh Graduate School of Public Health, Pittsburgh, PA., USA*
[b] *Irkutsk State Technical University, Irkutsk, RF*

Abstract. The Global Health Network Supercourse is assessed as a model for knowledge development and information sharing, and is considered to be new paradigm to reorganize all the global safety system. The difficulties which impede realization of the idea and implementation of networking are analyzed. Principles of InfoSec system building; dispelling myths; are implementing codes of practice described in ISO 17799 explored. And the interdisciplinary triangle of information security, library sciences and law is proposed to promote the idea of the network on a global scale.

Keywords. Networking, Prevention, Global Safety, Information Security

Introduction

Modern society is vulnerable to conventional threats from natural disasters such as hurricanes and infectious diseases to contemporary threats including nuclear waste contamination and terrorist attacks. While these threats can be overwhelming when taken as a whole, great strides have already been taken to address both conventional and contemporary threats: the invention of penicillin, tomography, satellite monitoring of the weather provide just a few examples. Currently, exponential progress is being made to address natural and man-made threats in one of the most versatile, powerful, and rapidly growing fields—information technology (IT). With the advent of computers, computer networks, the Internet, E-mail, the World Wide Web, and cell and satellite phones, we now have the tools to communicate beyond traditional social boundaries.

It is fair to say that the next phase in the information society will bring a boon of real-time E-business, E-government, E-management, remote E-working that will affect the majority of the population. These new technologies will further undermine the current hierarchical or vertical information sharing and data exchange. This traditional "silo" structure reflects the philosophy of how information, from financial data to personal information, is stored, organized, retrieved, and disseminated in Russia. Furthermore, this next phase will revolutionize how information can be shared by making local information available on a globally scale and visa versa. With that said, what is most important is the ability to create human networks.

1. Global Comprehensive Interdisciplinary Information Exchange

Global comprehensive interactive information exchange is necessary in today's world to effectively combat natural and man-made disasters. The groundwork has already been laid to promote such communication through the Global Health Network Supercourse [1]. Created in the 1990s at the University of Pittsburgh Graduate School of Public Health, one of the Disease Monitoring and Telecommunications World Health Organization (WHO) Collaborating Center, the Supercourse provides a model for global knowledge sharing and transfer, which can be used as model for human networks.

This library of electronic presentations, available internationally to anyone with an Internet connection or a computer that can run a CD-ROM, provides up-to-date multidisciplinary information in a universally accessible format. All stakeholders involved in public preparedness, disaster, and response from health educators to scientists can share information that will assist in advancing their countries' capacities. Creating knowledge and sharing information on a global scale through Internet networking technologies and through human networks is creating a new paradigm for a global security system. The question is, how receptive are countries like Russia to this new type of information network that levels the playing field in which information is shared?

2. Networking Models

Networking models can be divided into four main categories: conventional, interdisciplinary, inter-level, and international.

Conventional networking is the most basic form of networking. It's reliable and targeted, usually peer-to-peer, and takes place within an organization or discipline, between like-minded people.

Interdisciplinary networking is an effective, reliable, and productive information exchange among diverse disciplines. Examples of this can be seen in how experience was shared and implemented in the nuclear industry, the development of information security, and even in the information sharing that takes place on the Supercourse.

Inter-level networking is the sharing of information between different and diverse management and organizational levels such as: government to citizen, government to business, business to consumer. Most of this information regards problem-solving, prevention, and delegation of tasks, and often has a hierarchical structure.

International networking promotes the sharing and implementation of experience and expertise among nations.

3. Barriers to Networking

The advantages of implementing a human network can be seen in how the national economy was restructured, the creation of the modern ICT, as well as current European restructuring of the national economy, modern ICT, the experience in Europe and the United States, and in academia.

However, there are disadvantages that could impede the acceptance of this new concept of networking and its implementation. This can be seen especially in Russia where bureaucracy and inertia are barriers to any change in the hierarchical structure, along with skepticism toward relational connections throughout the world. Those who occupy positions of power employ an infrastructure that restricts the flow of information, so that the only sharing to take place is the sharing within that particular silo. Furthermore, even within this context some information might be restricted to a select few. In short, if national interests determine what information can and cannot be on to networks of people, attempting to access information including prevention of disasters, similar to information found on the Supercourse, would be almost a waste of time due to the incompleteness of the results.

In modern Russia, the obstacles and threats to changing how information flows is not a technological issue, but a social one. Because of the ingrained hierarchy and infrastructures built on families and family connections, many people are opposed to an egalitarian notion of knowledge. Here, the idea the knowledge is power means that many people want to restrict what information is available as well as to who it is available. Note that the separation of societies is created intentionally; security information is restricted, and business management is based mutual interests or on a family- friend interest structure.

4. Presenting the Supercourse as a Model for Promoting Social Networking

We aim to introduce and demonstrate the Supercourse as the nuclei of a new paradigm that can reorganize the entire global safety system, including social networking. We believe that Supercourse2, which is the next generation, will potentially assist us in persuading people to accept this new paradigm because it allows facilitation, collaboration, and cooperation, and can aid in improving global health in a cost-effective, secure, and user-friendly manner.

To date, the Supercourse has trained approximately 300,000 – 1 million students. Supercourse2 will provided enhanced systems for collaboration and information sharing; state-of-the-art technologies to build up the Supercourse in both numbers of public health participants, and ease-of-use. The Center of Excellence in Public Health Informatics, an organizational tool, include: creation of a system for creating, storing, accessing, integrating and manipulating information at various aggregate levels. This Extensible Public Health Information Collaboratory (EPIC) will allow synchronous and asynchronous collaboration among public health educators, and will provide an information infrastructure to reflect the knowledge and sophistication of the user, will create an environment that bridges cultural and linguistic boundaries, and will link geographic information systems to public health needs. This digital library will bring the ability to share information across the public health silos.

Personalization and real-time adaptation of the network can considered an essential issue in persuading people to accept the network of prevention paradigm. A personalized (or adaptive) system builds a model of the goals, preferences and knowledge of each individual user, and uses this model to guide interaction with the user. Not only the system should be adaptive, but users are to be prepared also: moving to meet each other. A personalized system helps to make users think collaboratively, to attract them into collaborative activity.

We regret, however, that the Center of Excellence employs more technological (even in linguistics terms) than information geared toward social aspects of networking. Even if some individuals or groups support the idea, it is necessary for people to accept the new paradigm of networking for the Supercourse ideals to spread. Another issue that prevents understanding and implementation of the new paradigm is security.

5. Information Security (InfoSec) Scope Supports New Paradigm for Networking

Obviously, networks and networking must be secure. It is widely accepted that “" . . . security has a value that can be broken down into three major attributes: confidentiality, integrity, and availability”[3]. Information must be readily available to those authorized to access it (availability), inaccessible to those not authorized to access it (confidentiality), and protected from destruction and/or unauthorized alteration of any kind (integrity). The system must meet the next seven principles of InfoSec System building, taking into account all possible and interconnecting threats including continuity, complexity, optimization, flexibility, simplicity, and openness.

5.1. Taking into Account All Possible and Interconnecting Threats

In the process of implementing the Supercourse into the global society, the team will encounter threats, which are different from those throughout conventional academic circles. Such threats will come from diverse groups which report contradictory information needs for, say, large and small corporations, the military, and universities.

Internal threats could be brought by a wide variety of individuals including researchers, practitioners, government officials, academics, military personnel, and corporate leaders. Each member of these groups has a different set of priorities, a different set of physical and technological resources, a different set of educational and information needs, and different codes of security practice.

Currently, in Russia, some groups take advantage of their positions by means of separation and anti-networking. In short, stakeholders might be more interested in protecting their interests or pushing their agenda without considering the larger benefit. Thus, network supporters might observe psychological, legislative, turf battles, even adversarial barriers to information retrieval and information dissemination as well.

5.2. Continuity

Creating a networking will be very different from the InfoSec process as it’s not just an a single and isolated act, rather maintaining its relevance demands continuous effort. The idea of updating of the Supercourse by contacting each of the authors of the lectures every two years and asking them for revised versions of their best lectures shows this very principle. The main obstacle is not just technological, but psychological as well—the ability to contact participants, and elicit responses.

5.3. Complexity

Complexity implies that all instruments from diverse disciplines must be applied to build the system. These instruments include the principal of prevention; ICT (including TV and PR); information security; law; library sciences. But first, we must dispel the myths that undermine our ability to apply the Infosec Scope to implementing codes of practice described in ISO 17799.

- **Myth 1: Networking in Academic Society is on appropriate stage.**
  Actually, academia needs networking as well as in any other discipline. The Supercourse is not a sole production on networking, still there are few. The Legal Framework for the Information Society (LEFIS) [4] demonstrates the academic approach to cross-disciplinary networking. Supported by the European Union as Thematic Network, LEFIS is the exemption rather than the rule in academic circles, and its experience might be thoroughly learned to further dissemination.
- **Myth 2: Networking policy is a technical issue.**
  Policies in networking are not technical manuals. Networking is about people.
- **Myth 3: If you build it, they will come.**
  Right after you post the new networking policy on the Internet, you notice all the users can think about is the extra work you have proposed to them. The process of writing policy is a burdensome group activity.

6. Codes of practice if follow ISO 17799 [5]

6.1. Terms and Definitions

The challenge to specialists of diverse disciplines is to design and develop effective means for collecting, representing, storing, retrieving, and disseminating relevant and useful information. This is especially true in any critical domain as public health in which accurate and useful information retrieved and delivered in the right form or mode at the right time and location to the right people can mean the difference between life and death. So it is of special value to work on a multi-language glossary of terms and definitions. From its origin the Supercourse has been developed as Multilanguage tool.

6.2. Implementation and Security Policy

One of the principal questions on InfoSec asks: Whose property is the information resources and processes? The classic answer may is the individual, corporation, or government. In some more complicated cases we observe collaborative project, and so individuals, corporations, governments try to defend their interests in the project. This is a persistent problem of the balance between secrecy and openness as the result of contradictions in ownership of information we see when universities, industries, and government interact.

7. Organizational Implementation and Security

Bringing users into the loop early makes them stakeholders in the policy and ambassadors to the rest of the networking sector. Peers who believe in the need for the Supercourse carry a lot more weight with users than the people who hold the Supercourse suspect. It is clear that public relations in academic and research fields differ from those of politics and trade but is of value as well.

7.1. Asset Classification and Control

All major information assets should be accounted for and have a nominated owner. Accountability for assets helps to ensure that appropriate trust and integrity is maintained. This can be accomplished using modern librarian technologies that apply contemporary ICT and have tools to identify the owners for all major assets.

7.2. Personnel Implementation and Security

It makes a great difference which people are engaged in the process—if there is a different team no doubt the Supercourse will be developed. But, as far as Thematic Network, LEFIS, is concerned, no doubt any other coordinator than Professor Fernanado Galindo leads that different way. Peers and users make difference as well.

Summary

Since Russia is a large and diverse country, it is an excellent laboratory in which it is possible to test the acceptance of the global system of the Supercourse, and with that a new paradigm for creating a social network of prevention. However, there are barriers to accepting this new way of networking among people. The obstacles are more social than technological. Nevertheless, we can be confident that with such resources we can provide roadmaps of the process, as seen in the example of the Information Security Techniques, with their ISO17799 ideology. Standards will help to build social networking to formulate information policy, clarify threats and barriers, name adversaries and competitors, to provide continuity. Additionally, librarian sciences and technologies will support dissemination of information severely and give it a solid base.

References

[1] Global Health Network Supercourse [Accessed April 10, 2006] Available from: http://www.pitt.edu/~super1/

[2] The National Strategy for the Physical Protection of Critical Infrastructures and Key Assets February, 2003.

[3] Balancing Value and Threats. [Accessed April 10, 2006] Available from: http://www.itillious.com/insight/articles/value_vs_threat.html

[4] Available from : http://www.lefis.org/index.htm.en Accessed July 26, 2005

[5] Information technology — Code of practice for information security management. *Technologies de l'information — Code de pratique pour la gestion de sécurité d'information.* International Standard ISO/IEC 17799. First edition 2000-12-01. Reference number ISO/IEC 17799:2000(E).

Scientific Networking and the Global Health Network Supercourse
F. Linkov and R. LaPorte (Eds.)
IOS Press, 2006
© 2006 IOS Press. All rights reserved.

Terrorism Preparedness and Evaluation in the United States: Are We Committed to Protecting All Segments of Our Population

Peter D. RUMM, MD, MPH[a], Jennifer SINIBALDI, MPH[a], Curtis CUMMINGS, MD, MPH[a]
Joe DIMINO, DO[b], Katie GLEASON, MPH[a], Holly WESTERGREN, BA[a]
[a] *Dr. Rumm, Dr. Cummings and Ms. Sinibaldi are at the Center for Public Readiness and Communications, Drexel University School of Public Health, Philadelphia, Pennsylvania (e-mail pdr26@drexel.edu)*
[b] *Dr. Dimino is the Health Officer and Director of the Montgomery County Health Department, Montgomery County, Pennsylvania*

Abstract: The United States has made some critical progress in beginning to measure the capabilities through the use of metrics and other evaluation tools. This action was called for by the lead author as part of the National Advisory Committee on Children and Terrorism after a lag period of relatively few evaluations except some capability instruments by the U.S. Centers for Disease Control and Prevention (CDC) and the Association of State and Territorial Health Officials (ASTHO). A recent review by Drexel University's Center for Public Health Readiness and Communication found some other evaluations with merit that correlated with national public health expert concern about preparedness. Recently, the CDC, as part of their larger U.S. cabinet, the Department of Health and Human Services (DHHS) began to work with the U.S. Department of Homeland Securities (DHS) National Response Plan experts and other collaborators to develop a set of critical capabilities that are part of a larger DHS initiative to measure Target Capabilities. Some public health experts in the U.S. do not see this trend towards capabilities over capacity as necessarily a great improvement, but it promises to track progress potentially more reliably and more often than in the past. On the other hand, it remains to be seen whether or not such Target Capabilities and more importantly, the anticipated decreased funding stream for "public health preparedness" will fully prepare some key sectors of our populations for various emergency situations

Keywords. Epidemiology, preparedness, public health, terrorism

Introduction

Are we measuring preparedness correctly? What is the current changing "climate" for completing this task in the United States? What is the impact of this seminal issue on our national and global security? The U.S. has spent approximately $4 billion on state and local preparedness in the past five years and Cohen and Sidell, among others, have questioned whether or not this funding improved public health. Others have responded that the U.S. public health system has dramatically improved while recognizing that years of relative neglect have left the system still under-funded and with some potential shortfalls [1; 2].This brief paper will primarily discuss some background history on this topic since September 11, 2001, through recent changes driven by the U.S. Department of Homeland Security and a draft National Disaster Response Plan led by that relatively new agency [3].

As background on their previous work in this area, Dr. Rumm is the director of Center for Public Health Readiness and Communication (CPHRC) at the Drexel School of Public Health in Philadelphia, Pennsylvania (Ms. Sinibaldi is the research coordinator). Two federal and state sponsored projects are underway at the CPHRC both aiming to develop metrics for regional preparedness. One of these projects is a rural and small urban initiative inside the Commonwealth of Pennsylvania referred to as the Lehigh Valley Project, and the second is a new tasking from two Regional U.S. Department of Health and Human Services (DHHS) administrators and the U.S. Assistant Secretary of Health to produce "metrics" for regional federal preparedness in conjunction with a collaborating partner Dr. Jack Thompson of the University of Washington's Center on Bioterrorism Preparedness in Seattle, Washington. Dr. Joe DiMino has published an article as a local health officer on the need for measures of emergency preparedness with his county (Montgomery County in Pennsylvania), and has worked to develop progressive and daily relationships with emergency response directors and other governmental agencies.

The NACCT and a Call for More Assessments

In 2003, Dr. Rumm served as the only state or local health official on the National Advisory Committee on Children and Terrorism [4]. This 12 member committee, which was supported by federal officials, was congressionally tasked (bill originally sponsored by Senator Hillary Clinton, NY) to produce a report on preparedness issues for children for the U.S. Secretary of Health and Human Services. One of the key recommendations of this report that came out months after the B. anthrax deaths in the United States called for more surveys and other evaluations on preparedness including that of children and other special populations.

The BTAIP and the CDC Capacity Instruments

Prior to this 2003 Report, there had been little formal release of the results of some capacity instruments for bioterrorism preparedness, especially those dealing with critical issues to the committee such as quarantine issues, legal issues, vaccine issues and emergency equipment concerns [5]. There were some lengthy surveys conducted on a volunteer basis for local and state health agencies produced by the U.S. Centers for DHHS. These were broadly termed "capacity" instruments and sought to measure static levels of preparedness.

Many states began to independently measure their capacity based on the current CDC funding into several areas: A) Emergency Response, B) Epidemiology, C) Laboratory Capacity, D) Chemical Laboratory Capacity, E) The Health Alert Network, F) Risk Communication, G) Education and Training. There was also separate funding from the Human Resources Services Administration (HRSA), another agency of DHHS for hospital preparedness. About the same time, the CDC, working with leaders of several public health groups and academic centers, crated a new inventory of what it considered critical preparedness capacity measures.

Subsequent information on this work by the CDC later became known as the CDC Critical Capacity Inventory. In the three years after September 11 and the U.S. anthrax attacks, this survey along with others was released by the U.S. Public Health Foundation (PHF) in 2004: "The Critical Capacity Inventory provides public health agencies with a rapid assessment of their ability to respond to public health threats and emergencies. The Capacity Inventory is organized into six chapters that correspond directly with the six funded focus Areas of the grant guidance—Preparedness Planning and Readiness Assessment, Surveillance and Epidemiology Capacity, Laboratory Capacity—Biologic Agents, Health Alert Network/Communications and Information Technology, Risk Communication and Health Information Dissemination, and Education and Training. Questions in the Capacity Inventory were derived in part from the Department of Justice Public Health Performance Assessment for Emergency Preparedness (i.e., DOJ survey), National Public Health Performance Standards assessment instruments, CDC Bioterrorism Core Capacity Project, Council of State and Territorial Epidemiologists (CSTE) Capacity Assessment Tool, and selected state specific assessment instruments" [6].

About the same time as the NACCT Report was formerly issued, the powerful policy body, the Association of State and Territorial Health Officers (ASTHO) produced a widely read report commonly referred to as the "Bioterrorism Accountability Inventory Project (BTAIP)" [7]. During that time, Dr. Rumm served as a chief medical officer for Wisconsin, and led a team of state officials and other contributors on local and state health preparedness.

A Recent Review of Major Assessments

At the Drexel School of Public Health CPHRC, Gaskins, Rumm, and other authors led a recent project to assess terrorism preparedness by a review of available local, state, and federal evaluations, surveys, and assessments in a work submitted to the *Journal of Homeland Security.* Gaskin and her co-authors found and reviewed in-depth approximately 50 such published evaluations. Some key findings of this draft work are:

- There is a paucity of published evaluations, especially in the area of children and special populations.
- Critical concerns commonly expressed include surge capacity for hospitals and various issues on communications systems (their reliability and sustainability).
- Command and control in emergency situation remains a major concern.

- In the minds of public health officials and other health care workers, major concerns remain about the capacity of the workforce to handle large epidemics, disasters, or acts of terrorism.
- Consistent with the ASTHO survey, laboratory capacity has improved as have several surveillance mechanisms. Some laboratories have syndromic surveillance capabilities.
- There have been important gains in vaccination planning and probably capabilities, i.e., through coordinating the National Defense Stockpile, the Public Health Readiness Initiative, and other efforts for major population centers [8].

Continuing Concerns by Public Health Officials

In a recent essay, as was mentioned earlier, a local health officer in Montgomery County, Pennsylvania, Dr. Dimino recently called for greater standardization of requirements, common methodology, and terminology [9]. Dr. Dimino's views stating a concern with local and state efforts seem to be the consensus of many other health leaders in the U.S. public health sector and from outside viewpoints at the local, state, and national levels.

One prominent expert in terrorism preparedness is Dr. Elin Gursky. Dr. Gursky, a former deputy health officer for New Jersey, has helped run a health care system, and then became a fellow in terrorism preparedness at Johns Hopkins University on the way to becoming senior policy leader for ANSER, a directorate for the U.S. DHS. Dr. Gursky, with this broad expertise, has been at times critical of the current inherent response capacity of local and state public health to respond to large-scale outbreaks or disasters, and has proposed, as one of several potential future public health response mechanisms, i.e., the creation of a "force protection" agency or force to respond at a federal level to major threats [10].

Such a force protection or more militarized/federalized format that may well be regional in structure in support of local needs may prove to be a necessary reality in the view of the authors as well. This might hold true especially if public health leaders themselves cannot figure out how to adapt quickly to the changing funding and political realities that are rapidly making them to "step up to the plate" and show effective preparedness for measurable outcomes. Dr. Gurksy suggests as well that as one part of this plausibility, such a format could potentially remove preparedness functions more directly from local and state health.

Recently, Dr. Shelley Hearne published a widely read report from Trust for America on the readiness of most U.S. states for terrorism preparedness. Many states scored very low on available metrics to measure preparedness [11].

Federal Oversight in Developing New Measures

As such, many experts in public health have recognized the need for more formal evaluation of preparedness including our federal agencies. Secretary Levett, the newly appointed Secretary of the U.S. DHHS and Julie Geberding, director of the CDC, have helped lead the development of a formal list of objectives in the current grant language for a new cycle pertaining to terrorism preparedness for public health and hospitals. The funding documents contain approximately three dozen such objectives [12].

These public federal efforts, to which the CPHRC has contributed, are part of a larger process by the U.S. DHS to develop Target Capabilities that deal with medical and public health from the National Disaster Response Plan. From recent DHS conferences, it is anticipated that a funding decision will follow nationally determined strategic goals and the needs to improve capabilities. It was also announced by Matt Mayer, an acting assistant secretary, that public health sector preparedness funding may be decreased overall, which was a point of marked concern by many state officials and other preparedness leaders. Areas that might be hardest hit are rural areas, smaller cities, and areas that perform in a substandard way on overall capabilities.

Capacity Versus Capabilities

Notice that there is a national shift in terminology in the area of evaluations and metrics for preparedness. The term "capabilities" is now being used and not "capacity." This changing terminology alarms some public health advocates, despite the fact that such shifts most likely signify a move towards measuring more in "real time" in order to test the validity of both current and improving capabilities.

At the recent ASTHO meeting, there was also a strong sense among some health officials that there could be a loss of an opportunity to continue to seek dual usage of badly needed funding to improve local and state health workforce and general disease control, as well as disaster and weapons of mass destruction preparedness [2]. This perception seemed especially acute and was verbalized vocally in an invited meeting of the U.S. State and Territorial Bioterrorism Directors (this was the first meeting of a newly formed group). Moreover, vocal opposition was raised about the use of the new DHS and DHHS target capabilities, goals, and objectives [13]. Recently, the RAND Corporation, under the direction of Dr. Nicole Lurie, has conducted a large federally funded (U.S. DHHS) project to evaluate local and state health preparedness through "real time" evaluations done in terms of exercises or blind phone calls to health officials on subjects like communication capabilities and command and control issues [14].

However, in the U.S. we live in times of tough fiscal decisions, and in many recent public forums federal officials have warned local, state, and academic partners to expect declining funding streams in the short- and long-term for preparedness [15]. Despite such concerns, it is highly likely that public health will have to continue to justify through assessment that we are improving preparedness through strict capabilities. We are rapidly moving toward strict measures, i.e., Does a health agency truly have intact 24/7 on-call capability at all times and more importantly, does it stand up to unannounced testing?

Conclusions and Continued Concerns about Readiness, Measurements, and Funding

From the start, it is imperative that we accurately conduct this measurement using the proper metrics and objectives. By not doing so, we place our nation's preparedness, as well as our population's safety, including that of children and special needs groups at risk. Do not forget that public health was built on the three core principles: 1) Assessment; 2) Assurance; and 3) Policy Development [1]. So, if we can measure it we ought to arguably be able to drive policies to assure and improve preparedness, or at least make a stronger case for continued funding and maintaining the national will to improve public health preparedness.

We remain somewhat optimistic that our nation is on the right track by moving toward more rigorous measurement and the use of target capabilities. However, there is no doubt that public health also needs to continue to build capacity after decades of decreased relative funding versus other medical spending.

In fact, in 2004 the PHF estimated that only 6.4 percent of all health care spending is attributed to public health. Despite an input of approximately 2.8 billion from the CDC alone to states in the last three fiscal years, the PHF wonders if this funding has supplemented other needs and has improved overall public health readiness.

Current public health spending may include considerable amounts of funding designated for homeland security efforts. It is too early to know the effectiveness of these expenditures on the health of the overall population [17].

As such, we would like to see such work contain specific objectives to ensure the preparedness of children, the aged, and the disabled and other special populations. In addition, the results should be shared and evaluated to ensure that we are measuring various aspects of disasters and preparedness "the right way," while making progress within the bounds of national security concerns.

References

[1] Cohen H, Gold R, and Sidell V. 2004 Oct. The pitfalls of bioterrorism preparedness: the anthrax and smallpox experiences. Am J Public Health 94(10):1667-71.

[2] Rumm PD. 2005 March. Bioterrorism preparedness: potential threats remain. Am J Public Health 95(3):372.

[3] National Response Plan (Draft). U.S. Department of Homeland Security, produced Oct. 2004 for comment. Available from: http://www.dhs.gov/dhspublic/interapp/editorial/editorial_0566.xml

[4] Recommendations to the Secretary, Report of the National Advisory Committee on Children and Terrorism, DHHS. Released 2003. Available from: http://www.bt.cdc.gov/children/recommend.asp

[5] Diaz A, Rumm P et. al. 2004. National advisory committee on children and terrorism. Schools and Terrorism: A Supplement to the Report of the National Advisory Committee on Children and Terrorism. Journal of School Health 74(2): 39-51.

[6] Public Health Foundation. Available from: http://www.phf.org/infrastructure/phfpage.php?page_id=43&pp_id=19 and linked to original CDC site available from: http://www.phppo.cdc.gov/od/inventory/index.asp

[7] ASTHO. Bioterrorism Accountability Indicators (BTAIP). (Project, released 2003) Available from: http://www.astho.org/templates/display_pub.php?pub_id=643

[8] Gaskins M., Rumm P, Cummings C, and Hu X. 2005. An update of terrorism preparedness evaluations, two years after the Bioterroism Preparedness Accountability Indicators Project. Submitted to the *Journal of Homeland Security* and its abstract has been accepted by the American Public Health Association for presentation at 2005 meeting.

[9] Dimino J. 2004 February. Public Health and Homeland Security Are We Ready? ASTM Standardization News.

[10] Gursky E. 2004. Drafted to Fight Terrorism: Public Health on the Front Lines of Biodefense, ANSER.

[11] Hearne S. 2005. Trust for America – Protecting America's Health in the Age of Bioterrorism.

[12] 2005-06 CDC and HRSA/DHS funding cycle documents for state and territorial preparedness. Available from: http://www.bt.cdc.gov

[13] Inaugural Meeting of the State and Territorial Bioterrorism and Preparedness Directors by ASTHO, Boston, USA 2005 July 11-23.

[14] Personal communication and meeting with Dr. Lurie by Dr. Rumm and Ms. Sinibaldi, RAND Corporation, Pentagon City, Virginia. 2005 July.

[15] Various statements on expected funding by federal and state officials at meeting including: The ACPM meeting in DC in Feb 2005; The U.S. Public Health Officers Association Meeting in Philadelphia in June 2005; The CSTE Meeting in New Mexico in June 2005; and The first Annual Meeting of State and Territorial Bioterrorism Directors and the combined ASTHO/NACCHO Meeting (both held in Boston in July 2005).
[16] Turnock B. Pubic Health –What It Is and How it Works. Aspen Press, 2004 (3rd ed.).
[17] U.S. Public Health Foundation Web site on Percent of Health Dollars on Public Health. Available from: http://www.unitedhealthfoundation.org/shr2004/components/percentph.html

Scientific Networking and the Global Health Network Supercourse
F. Linkov and R. LaPorte (Eds.)
IOS Press, 2006

© 2006 IOS Press. All rights reserved.

Globalization of Information Sharing and Modern Technology

Rania Ahmed AZMI
Faculty of Commerce, Alexandria University, Egypt

Abstract. Globalization and technology present challenges and opportunities for both new and existing channels of scientific networking. Scientific research activities can be considered a foundation of cooperative work in a worldwide context. In this regard, joint research projects and research networks are enabling structures for scientific collaboration that is critical for developed and developing countries. The idea of scientific networking is beneficial for both the individual and the community, for example, when the expertise, the skills, and the vision of different members of the group can be shared by the community in a particular country or even by other countries particularly between developed and developing countries. Information can be turned into knowledge and, therefore, information sharing can lead to knowledge sharing. But knowledge sharing—which is the core of scientific collaboration—is a process that embeds the notion of the willingness to share. Cooperation can be defined as the work in progress, bringing the specialties of different participants that complement each other to the network. With the advent of Internet, the networking function is very easily accomplished. What is then important is that each scientist tries to establish a cooperative database that is globally accessible or at least tries to provide his/her scientific contribution to pre-established databases.

Keywords. Globalization, information, public health, technology

"Science is a global activity and will be advanced by our cooperation."
—George Strawn

Introduction

There is a need to create an environment in which all the personal scientific data or information assets of an individual or an organization can be available for the benefit of the scientific community within one country or across the globe.

No one can deny the growing importance of computer networks to scientific research and the importance of the Internet in the area of scientific networking. This is demonstrated by the fact that medical work, for instance, involves a team of multidisciplinary people working together to care for their patients with medical information systems needed to support collaboration among these individuals. Furthermore, such collaboration is needed for helping the underprivileged in developing countries to access the critical medical information databases. Without such networking available for the benefit of the underprivileged in developing countries, the digital divide would probably take place separating the "haves from the have-nots." The digital divide means that among countries or among different groups of people within one country, there is a wide division between those who have real access to information (whether medical, financial, agricultural, or industrial information, etc.) and communication technology, and are using it effectively from those who do not.

Global scientific collaboration and networking offer the best opportunity to bridge the digital divide in a sustained and effective manner. The Global Health Network Supercourse project <www.pitt.edu/~super1> is one of the largest efforts in scientific networking in the area of public health. The Supercourse project is a global library of lectures on prevention and the network of scientists that is gaining increasing popularity in Egypt. The goal of this paper is to provide an in-depth analysis of the issue of the globalization of information sharing and the role of modern technology in this regard, particularly for the purpose of scientific collaboration and networking.

Globalization and technology are transforming the use of information in a rapid manner. This is demonstrated in a worldwide context by scientific communications and collaborations. With the advent of Internet, the networking function is very easily accomplished, but further steps should be taken to enhance scientific networking and collaboration to extend to new fields or even to new places. Part two of this paper introduces basic concepts of the globalization of information sharing and its role in scientific networking and collaboration. Part three briefly reviews the common context in which technology and electronic science (E-Science) are discussed. And, finally, part four is devoted to the concluding remarks and implications, particularly for scientific networking.

Globalization of Information Sharing

The information society is an evolving phenomenon that has reached different levels across the world reflecting different stages of development. Globalization is rapidly transforming the environment in which the information society is developed. Therefore, globalization has led to rapid information sharing between countries [1].

As research becomes increasingly global there is a growing need to address information access and data sharing issues. Access and sharing of data/information are essential for the conduct and advancement of sciences. Following are brief definitions of the terms "access" and "sharing" in the context of collaboration and networking [2]:

- Access is the act of making data/information available for use by others
- Sharing is a researcher sharing data/information one or more individuals

The sharing of data/information involves providing specific access, whereas the act of providing access by itself does not necessarily involve any sharing arrangement.

In an era of global collaborations and data intensive grids, advanced networks are required to seamlessly interconnect physics research groups enabling them to collaborate throughout the lifecycle of their work. Ensuring research data/information are easily accessible so that they can be used as often and as widely as possible is a matter of sound stewardship of public recourses of researches [3]. Therefore, there is a need to create an environment in which all the scientific data/information assets of an individual or an organization can be harnessed for the benefit of the large scientific community [4].

Received by the right individual, someone who understands its meaning and possibilities, and in the adequate delta of time, information can be turned into knowledge. In this case, information sharing can lead to knowledge sharing. Knowledge is collectively owned and maintained by the community in the context of broad scientific collaboration, therefore, the motivations for knowledge exchange is not self-interest, but are for the community [5, 6].

Technology and E-Science

Technology has changed the way information and research is produced, reviewed, published, accessed, and used. This change presents challenges and opportunities for both new and existing channels of scientific communication [7].

In the context of modern technology, the major drawbacks of traditional scientific research methods include the following: First, scientific research activities have been basically closed or the extent of information exchange has been far from enough. As a result, historically, it often occurred in that many different individuals achieved the same observations separately and almost simultaneously. If information exchange had been more adequate, much unnecessary and repetitive labor would have been saved and the time to make breakthroughs would have been advanced greatly. Second, the means of scientific research activities have been subjected to a lot of limitations, and especially to a lack of simulating and emulating capability. So, traditional scientific research activities are often characterized by a long cycle and high cost, and many experiments are even impossible to do at all. All these limitations have restricted the development of thinking and collaboration while the emergence and rapid development of computer technology has radically changed such situation [8, 9].

With the advent of the Internet, the networking function is very easily accomplished. What is then important is that each scientist or scientific community or organization develops their own databases accessible through the Web for the global scientific community [10].

Computer networks are critical to scientific research collaboration. They are currently being used by portions of the scientific community to support access to information resources and collaborative work through such facilities as E-mail and shared databases. There is considerable movement in the direction of providing these capabilities to the broad scientific community in a unified manner [11, 12].

Scientists were among the first to recognize the potential of emerging information and communication technology (ICT) for enhancing and extending their work and for globalizing their research. For example, electronic mail first became widespread within scientific communities [13, 14]. Therefore, one of the most important areas of information sharing and communication via modern technology is E-science, briefly defined as increased access via desktop or other interface via the Internet to distribute resources and global collaboration [15; 12].

High-speed computer networks grew and connected increasingly large numbers of researchers, who could share information. Now the modern world requires effective information sharing across countries through the Internet [16; 17]. The network will build powerful E-science resources for the structural sciences and for greater collaborations. E-science provides open access to the digital output of scientific research resulting in a wide collaboration and networking of scientific information [18; 19; 20].

Conclusion

Globalization and technology play a central role in scientific collaboration worldwide. Associating people, institutions, or organizations in networks is beneficial for both the individual and the community. Still a work in progress, scientific collaboration is bringing together people who complement each other's work into the scientific community through the network. In short, globalization and modern technological innovation reinforce each other; they are two critical forces driving the networking of scientific information and collaboration toward realizing the potential for fruitful communication within a global context.

References

[1] Iamnitchi A, Ripeanu M, and Foster I. 2005. Locating data in a small world? Peer-to-peer scientific collaborations. Department of Computer Science, University of Chicago.

[2] Rycroft R. 2002. Technology-based globalization indicators: the centrality of innovation network data. Occasional Paper Series.

[3] Mujahid Y. 2002. Digital opportunity initiative for Pakistan. COMSATS Internet Services, Islamabad.

[4] Davidson E. and Lamb R. 2004. Examining socio-technical networks in scientific academia. Industry Collaborations. University of Hawaii;1625-1631 p.

[5] Pratt W, Reddy M, McDonald D, Hornoch P, and Gennari J. 2004. Incorporating ideas from computer-supported cooperative work. Journal of Biomedical Informatics 37:128-137.

[6] Nabuco O, Rosario J, Silva J, and Drira K. 2002. Scientific collaboration and knowledge sharing in the virtual manufacturing network;. IFAC.
Available from: http://www.pmr.poli.usp.br/d-lab/artigos/140.pdf

[7] Grudin Jonathan. 2004. Crossing the divide. ACM Transactions on Computer-Human Interaction 11(1):1-25 p.

[8] Hu, Q. 2005. The open access to scientific information in China. Chinese Association for S&T.

[9] Mianheng, J. 2002. Information of Scientific Research and E-Science; The Chinese Academy of Sciences. Shanghai. 2002.

[10] Warnasuriya, Dilmani. 2001.Interchange and exchange of scientific information among S&T information centers. Sri Lanka Country Paper.

[11] Gennari J, Silberfien A and Wiley J. 2005. Integrating genomic knowledge sources through anatomy ontology. University of Washington: Biomedical and Health Informatics. Information School, and Comparative Medicine.

[12] Ali A. 2003. Benefits of International Scientific Collaborations for Developing Countries. Round Table on Developing Countries Access to Scientific Knowledge, The Abdus Salam ICTP, Trieste, Italy. 97-100 p.

[13] Houghton J, Steele C, and Henty M. 2003. Changing research practices in the digital information and communication environment. Australia: Department of Education, Science and Training.

[14] David 2003. The digital technology boomerang: new intellectual property rights threaten global open science.World Bank Conference; ABCDE.1-32 p.

[15] Agouris P, Cresswell T, Hovy E, Katzy B, and Macintosh A. 2004. International collaboration on E-government research. A white paper established at a joint meeting between the EU and US delegations at the NSF's Fifth National Digital Government Conference. 1-21p.

[16] Arzberger P, Schroeder P, Beaulieu A, Bowker G, Casey K, Laaksonen L, Moorman D, Uhlir P, and Wouters P. 2004 November 29. Promoting access to public research data for scientific, economic and social development. Data Science Journal 3:135-152.

[17] Castell M. 1999. Information technology, globalization and social development. UNRISD Discussion Paper No. 114, United Nations Research Institute for Social Development.

[18] Jebaraj F. and Devadoss F. 2004. Library and information networks in India. Dept. of Library and Information Science, Madurari Kamaraj, University, India.Library Philosophy and Practice 6(2).

[19] David PA. 2001. Will building 'good fences' really make 'good neighbors' in science? The consequences for Internet-mediated research collaborations of broadening IPR protections, European Commission (DG-Research). Available from: http://europa.eu.int/comm./research/headlines/archive.html#03

[20] Leiner B. 1987. Network requirements for scientific research: Internet task force on scientific computing. Network Working Group;1-19 p.
[21] Olson G, Teasley S, Bietz M, and Cogburn D. 2002. Collaboratories to support distributed science: the example of international HIV/AIDS research proceedings of SAICSIT; 44-51 p.
[22] Saxena A, Papitashvili S. 2004. A peer-to-peer semantic approach for efficient sharing of scientific data. Michigan: Department of Electrical Engineering and Computer Michigan.
[23] Yusuf S. 2001.Globalization and the Challenge for Developing Countries. DECRG, World Bank.

Methodologies for Preparedness for Disaster Situations: Examples from NATO Countries

Scientific Networking and the Global Health Network Supercourse
F. Linkov and R. LaPorte (Eds.)
IOS Press, 2006

© 2006 IOS Press. All rights reserved.

Disaster Database: A New Experimental Model

Alessandra ROSSODIVITA, MD
Department of Cardiothoracic and Vascular Diseases
Member of Maxi-Emergency Committee
San Raffaele Foundation

Abstract. Disasters are generally considered "low probability–high impact events." Disasters always arise from some disequilibrium between hazards in the environment and the vulnerabilities of human communities. In the 21st century natural and complex disasters caused a dramatic increase in the demand for emergency medical care. Epidemiological data are well recognized as essential to understand to competent disaster management. Studies of disaster impact on public health have suffered by the lack of these data, especially the lack of real time .The analysis of hazards and vulnerabilities represent the key to managing and mitigate disasters assessment. The objective of this project is to create a country database (DB), an Italian Database, to collect all available information about the past country disasters and to establish a model for future, with a an analysis of factors that are most likely to produce hazards. The aim of this model is to investigate the impact of disasters upon the physical security and the health of people; factors that are most likely to produce hazards.

Keywords: Disaster, disaster epidemiology, injury

"Virtue is bold, and goodness never fearful." —William Shakespeare, *Measure for Measure*, Act III

Introduction

In the 21st century natural and complex disasters caused a dramatic increase in the demand for emergency medical care. Ability to predict future disasters could drive the efforts of the public policy and government funding. To understand the next century of disasters we must understand the concept of disaster. Disaster is a multi-various concept and needs to have a precise definition. According to the World Health Organization (WHO), a disaster is "a sudden ecological phenomenon of sufficient magnitude to require external assistance" [1]. At the community level, this can be defined as any community emergency that seriously affects people's lives and property and exceeds the capacity of the community to respond effectively that emergency [2; 3]. Gunn suggests that a disaster results from "a vast ecological breakdown in the relations between man and their environment, a serious and sudden event on such scale that the stricken community needs extraordinary efforts to cope with it, often with outside help or international aid" [4]. Disasters are generally considered "low probability–high impact events." This suggests that disasters always arise from some disequilibrium between hazards in the environment and the vulnerabilities of human communities [5]. The analysis of hazards and vulnerabilities represent the key to managing, preventing, and mitigating disasters.

The factors tending to produce hazards in the 21st century include population growth, environmental degradation, infectious agents, hazardous materials, economic imbalance, global warming, conflicts, and terrorism. During the last decade, terrorism-related emergencies and conflicts have represented a new resurgence involving a complex pattern of global changes and imbalances [6]. Throughout the globe, healthcare providers are increasingly challenged with the spectre of terrorism and the fallout from weapons of mass destruction [7].

The risk of any given disaster is modifiable through its manageability. Efforts to establish better preparedness for and prevention of disasters have been a priority of many countries. A good disaster management plan requires accurate information and must link data collection and analysis to an immediate decision-making process. Epidemiological data are well recognized as essential to understand and to manage a disaster to allow a progressive improvement in the ability of the health system to respond more effectively and efficiently to disasters. Studies of disaster impact on public health have been lacking such data, especially due to lack of time. Furthermore, we suffer from the lack of a systematically collected data. The analysis of hazards and vulnerabilities represent the key to managing and mitigation of disasters assessment. Planners, policymakers, authorities, and all other forces engaged in emergency management have all expressed need for these data [8; 9; 10].

1. Aim of this Study

The aim of this study is to create a national database, an Italian database (ITEM-DAT= Italian emergency database) for collecting all available data about historical information on disasters that have directly affected Italians. Additionally, the aim is to establish a model for the future with an analysis of the factors that are most likely to produce hazards, and to allow for the collection of so-called "perishable data," and all the data previously not available due to a "lack access" (as so-called confidential data or under proprietary rights).The database will describe where and when a disaster occurred, who was affected, and will attempt to provide an estimate of the direct costs. This database will contain references to all types of Italian disaster situations including those triggered by natural hazards, anthropogenic disasters (technological hazards, terrorism and conflict related hazards), and vulnerabilities (population growth, population aging, poverty, urbanization, population migration, and humanitarian emergencies related to the complex humanitarian emergencies and refugees).

The aim of this study is to improve the sharing of public health information, to improve and to balance the Italian Health and Security scenario in all Italian districts related to a disaster in terms of preparedness, as well as in mitigation and prevention programmes to rationalize the decision-making process for disaster preparedness and management, and to serve the purpose of humanitarian action at national and international levels.

Actually, there were many international databases created, with varying criteria for formats and purpose, in response to the need for better data on disaster occurrence [11; 12; 13; 14; 15]. The ITEM-DAT developed with the aim to create a Model Constitution of a National Working Group for the Study of Disaster Epidemiology (ITEM – Italian Emergency) linked to the establishment of a National Database (ITEM-DAT) and a World Wide Web site.

The National Working Group for the Study of Disaster Epidemiology and the establishment of a National Database (ITEM Working Group and ITEM-DAT) will be a permanent working group of researchers and practitioners who are competent to investigate and study the impact of disasters upon the physical security and the health of people. The group will consist of experts from different fields of medicine, public health, epidemiology, toxicology, volcanology, seismology, geography, engineering, emergency planning and management, urban planning, sociology, fire brigades, civil protection, anthropology, economics, the Red Cross, NGOs, volunteers groups, and any discipline that deals with the impact of natural or anthropogenic disasters on security and health of human populations .

1.1 Relationship of the Group with Other Relevant Organizations

This group will be organized and administered by members of universities and professions in the public and private sectors. It will be independent of the control of the state, public, or private institutions, but with the objective to maintain a strict collaboration with other relevant organisations at the national and international levels. These organizations include the National Health Service, national technique services, the World Seismic Safety Initiative (WSSI), the Centers for Disease Control and Prevention (CDC), the World Health Organization (WHO), and the International Association of Earthquake Engineering (IAEE).

1.2 Scope of the Group's Activities

Members of this ITEM-working group will seek to make it a significant element of any post-disaster investigation conducted in the country or as a component of investigations abroad. The main objective will be to study the effects on human health from catastrophic natural and anthropogenic phenomena such as earthquake, volcanic eruptions, floods, tornadoes tropical cyclones, landslides, transportation crashes, industrial explosions, and toxic chemical releases, with respect to both the prediction phase and long-term aftermath of disaster. Within this group, there will be a special task force group focusing on analyzing terrorism data and its public health consequences.

1.3 Applied Focus

This network will concentrate its work on applied science and on methods of planning in order to contribute to the reduction in the loss of human lives, injuries, and suffering associated with disasters. This initiative will require:

- A continuous improvement of methods used to estimate the numbers, types, and location of injuries caused by a disaster;
- The study of the means of reducing vulnerability to injuries during a disaster;
- The use of research results to improve the policies for the reduction of risks and mitigation of catastrophe;
- A national studying program of training an education to increase the level of knowledge about disasters; there is also a possibility to create an E-book on disasters. This book would be developed in user friendly format and an updated disaster glossary;

- International research of links related to disasters as well as the collection of papers and scripts from all over the world to share the international current position on disasters;
- The establishment of a series of national and regional meetings at which the results of the research activities in disaster epidemiology will be presented and discussed;
- The group will also seek to publish the results;
- Each member will be encouraged to keep in contact with each other, in addition a national and international register will be compiled in order to contain a list of the names and contact details of experts in disaster epidemiology.

2. Aim of the Group

The ITEM-working group will study injuries to human beings caused by disasters. These will be evaluated in relation to:

- mortality and physical injuries caused directly by the impact of disasters
- illnesses and mental disturbances resulting from such impacts
- the physiological impact of mourning
- any actual and potential increase of disease transmission and the public health consequences
- other illness caused directly or indirectly by disasters
- nuclear, biological, and chemical (NBC) terrorism and conflict related disasters
- ethical considerations

The ITEM-group will study each phase of the disaster cycle and interactions:

- before the event (the prediction of the context)
- the emergency phase (the context of rapid reaction)
- the short-to-medium term aftermath (the context of recovery)
- the long term (the context of persistent effects)

The ITEM-group will conduct risk analysis:

- risk and impacts at individual level
- situations of collective risk or impact at wide scales
- risks and impacts referred to particular buildings, sites, situations and places

The ITEM-group will study different forms of hazards—an analysis of the human health impacts related to a disaster respect to the following types of hazards:

- collapse of buildings and its connotation in terms of mortality and morbidity
- structural mortality and morbidity caused by secondary impact such as seismic landslides, tsunamis, and post-earthquake fires
- morality and morbidity related to an increase in the level of disease transmission, considering the new, emerging or remerging infectious diseases
- mortality and morbidity caused by NBC and terrorism-related injuries, and all causal factors that are unpredictable or uncontrollable
- studies on hazardous materials, caused by industrial chemical spills, gas leaks, industrial explosions and fires

The ITEM-groups will study specific tasks:

- improve the comparability and compatibility of data

- definitions of a standard common methodology to collect each data and procedures related
- a theoretical basis for the study of disaster epidemiology on the basis of the international current concepts
- a World Wide Web Site (under construction) to display and share all available information with the main objective to facilitate rapid communication among different researchers, which will allow the collection of so-called "perishable" data (e.g., the exact number of casualties after an earthquake, or after a tsunamis) and all data related to "a lack of access" because its considered confidential or under proprietary rights. These data will form a centralised archive on disaster injury epidemiology, easy to find and understand. The aim of this project is to improve and to balance the Italian health and security scenario in all districts, related to disaster situations.

3. Database and Basic Principles

The purpose of an ITEM-DAT is to rationalise the decision making process for disaster preparedness, as well as to allow for obtaining accurate and valid information about disasters; the main objective is to outline the importance of maintaining disasters records. Feedback on the Italian disaster database situation indicates that government agencies, researchers, educational institutions, emergencies service agencies, and all forces engaged in emergency suffer from lack of data collection that arises from a jeopardized educational program, and a past not well-recognized need to establish a common nationally consistent system of data collection. The growing numbers of disasters requiring external assistance, and the international global situation, has highlighted a new interest in mutual collaboration among all forces engaged in emergency in terms of predicting future disasters, as well as in mitigation, and a prevention program, drives the priorities and urgencies of our nation. The ITEM-DAT serves the purpose of humanitarian action at the national and international levels. Actually, these databases will not replace other international databases. This database is generally limited to Italian scenario, and has not been compatible with other existing databases.

3.1 Database Structure

The database structure is continuously being updated to allow finding all data available. The database will analyze all Italian scenarios, with an analysis of Italian territory divided in 20 districts (Figure 1). Italy will be divided into districts and the name of the region in which disaster occurs will be entered directly. The study of some geographical maps linked to the agricultural, industrial, environmental, and population aspects allow recognition of emerging situations. The database will be divided into three different sections:

- Catastrophic natural—earthquakes, floods, volcanic eruption, storms landslides, avalanche, infectious agents (Table 1.)
- Anthropogenic hazards—transportation crashes, hazardous materials, industrial explosions and toxic and chemical releases, nuclear materials, biological warfare agents, terrorism, conflict related (Table 2 and 3.)

- Vulnerabilities—population growth, population aging, poverty, urbanization, population misdistribution, complex humanitarian emergencies as refugees and immigration (Table 4.) [16; 17; 18; 19; 20; 21; 22; 23; 24; 25; 26].

There will be five numeric variables that require exact values: (1) people killed (2) people affected (3) homeless (4) injured and (5) the monetary cost estimation or losses sustained:

1. Number killed: persons confirmed dead and persons missing or presumed dead.
2. Number of people of affected: people requiring immediate assistance during an emerging situation; they are always part of the "total affected" population; Total affected: is the sum of injured, affected and homeless people.
3. Number of homeless: as people needing immediate assistance with shelter; they are part of the total affected population.
4. Number of injured: people with physical injuries/trauma/illness requiring medical treatment as a direct result of disasters. The injured are always part of the "total affected."
5. Total estimated damages: in terms of economic losses or damage directly or indirectly related to the disaster, a value in Euros or U.S. dollars.

4. Final Considerations

Disasters affect the community in numerous ways. One of the main problems in a disaster situation is that it remains hypothetical until it manifests as an event. It is crucial to create a national network with the objective to analyze, prevent, and mitigate disasters in terms of better communication in the area of public health. Building an Italian database and network for disasters and prevention would produce a sharing of information for an emergency situation to assess the needs of a disaster affected populations, find all available resources to those needs, prevent further adverse health effects, implement disease control strategies, establish a training program to evaluate the effectiveness of disaster relief programs, improve plans for various type of future disasters, and to create a real-time public health surveillance system as well an educational health emergency program in all Italian districts .

This paper represents an attempt to show how to collect information about the nature of various types of disasters and the associated working group as a template for other countries to compare and to share all these data. We have the responsibility to apply these data and this knowledge to preserve humanity.

Table 1. Natural Disasters

Natural Disaster type	**Disaster sub-type**
Drought	
Earthquake	
Epidemic	Antrax Arbovirus Diarrhoeal Diptheria Intestinal Protozoal Leptospirosis Measles Meningitis Plague Rabies Respiratory Rickettsiosis Small Pox Viral Epatitis Avian Influence
Extreme temperature	Cold Wave Heat Wave
Flood	Flash food Coastal flood Plain flood Valley flood
Insect infestation	Grasshopper Locust Other
Slide	Avalanche Landslide Mudflow
Volcano	
Wave/surge	Tsunami Tidal Wave
Wildfire	Forest Peat Scrub Steppe
Windstorm	Cyclone Hurricane Tornado Tropical storm Storm Typhoon Winter storm
Objects from outer space	Meterorite Satellite Technological Objects

Table 2. Anthropogenic Hazards

Anthropogenic hazards type	**Disaster subtype**
Industrial accident	Chemical spills Collapse Explosion Fire Gas leak Poisoning Radiation Other
Transport accident	Air Rail Road Water
Miscellaneous accident	Collapsed buildings Fire Urban chemical fire with chemical spills Blackout Other

Table 3. Terrorism Hazards

Terrorism hazards	**Subtype**	**Subtype**
Explosion	Blast explosion conventional Blast explosion non-conventional	Chemical or nuclear
Biological agents	Cat A	Antrax Clostrid. botulinum Yersinia Pestis Variola major(smallpox) Pox virus (other) Francisella tularensis Viral hemorrhagic fevers Arenaviruses LCM, Junin virus, Machupo virus; Guanarito virus; Lassa fever; Bunyaviruses (Hantaviruses, Rift valley fever) Flaviruses (Dengue, Filoviruses, Ebola, Marburg)

Table 3.1 Terrorism Hazards

Biological Agents	Cat B	Coxiella burnetii (Q fever) Brucellosis Ricin toxin Epsilon toxin of Clostrid.Perfrigens Stafilococco (enterotoxin B) Rickettsia prowazekii (typhus fever) Food and waterborne pathogens **Bacteria :** E.coli Shigella Salmonella Listeria monocit. Campylobacter jejuni Yersinia Enterocolitica **Viruses :** Hepatitis A Caliciviruses **Protozoa :** Cryptosporidium parvum Cyclospora cayatanensis Guardia lamblia Entamoeba histolytica Toxoplasma Microsporidia **Additional viral encephalitides:** West Nile Virus La Crosse California enceph VEE EEE WEE Japanese encephal. Viruse Kyasanur Forest Virus

Table 4. Vulnerabilities

Population growth	Years Sex Age
Population distribution	Rural Urbanisation
Refugees	National Extra-national
Poverty	Analysis of Different districts

Figure 1. Italian Districts

References

[1] Noji EK. 1997. The public health consequences of disasters. New York (NY): Oxford University Press.
[2] Pan-American Health Organization: Emergency Health Management after Natural Disaster. Scientific publication. 1981.no.407, Washington (DC): PAHO Office of Emergency Preparedness and Disaster Relief Coordination.
[3] World Health Organization: Emergency care in natural disasters: views of an international seminar. 1980. Chron 34: 96.
[4] Gunn SW. Multilingual dictionary of disaster medicine and international relief. Prehospital and Disaster Medicine Web site. [Accessed March 2002] Available from: http:// pdm.medicine.wis.edu/vocab.htm
[5] Arnold JL. 2002. Disaster Medicine in the 21st Century: Future hazards, vulnerabilities, and risk. Prehosp Disast Med 17(1):3-11.
[6] Alexander D. 2003. Terrorism, disasters, and security. Prehosp Disast Med 18(3):65-169.
[7] Larkin GL, Arnold J. 2003. Ethical considerations in emergency planning, preparedness and response to acts of terrorism. Prehosp Disast Med 18(3):170-178.
[8] Noji EK.2005. Public health issues in disasters Critical Care Med 33(1):29-33.
[9] Noji EK. 2002. Field investigations of natural disasters. Field Epidemiology. Gregg MB (Ed).New York (NY): Oxford University Press. 365-383 p.
[10] Hogan DE, Burnstein JL 2002. Disaster Medicine. Lippincott Williams & Wilkins.
[11] EM-DAT Emergency Disasters Data–Base. Available from: http:// www. em-dat.net
[12] Australian Emergency Database EMA Available from http://www.EMA.gov.au/
[13] Canadian Disaster Database. Available from: http://www.ocipep.gc.ca/disaster/default.asp
[14] Natural Disaster Reference Database. Available from: http://www.ndrd.gsfc.nasa.gov/ndrd/
[15] CRED-Center for Research of Epidemiology Disaster. Available from: http: //www.cred.be/site.map.htm

[16] Natural Disasters & Severe Weather. Chemical Emergencies. Centers for Disease Control and Prevention. Available from: at http:// www.cdc.gov/
[17] Mass Casualties. Centers for Disease Control and Prevention. Available from:http://www.cdc.gov/
[18] Bioterrorism Agents. Centers for Disease Control and Prevention. Available from: http:// www.cdc.gov/
[19] Radiation Emergency. Centers for Disease Control and Prevention. Available from: http:// www.cdc.gov
[20] Larkin GL, Arnold J. 2003. Ethical considerations in emergency planning, preparedness and response to acts of terrorism. Prehosp Disast Med 18(3):170-178.
[21] Hien TT, De Jong M, Farrar J. 2004. Avian influenza-a challenge to global health care. N Engl J Med 351(23): 2363-2365.
[22] Osterholm MT. 2005. Preparing for the next pandemic N Engl J Med 352 (18) 1839-1842.
[23] De Palma RG, and others 2005. Current Concepts : Blast Injuries. N Engl J Med 352 (13) 1335-1342.
[24] Aas P. 2003. Future considerations for the medical management of nerve agent intoxication. Prehosp Disast Med 18(3):208-216.
[25] Di Giovanni C Jr. 2003. The spectrum of human reactions to terrorist attacks with weapons of mass destruction: early management considerations. Prehosp Disast Medi 18(3)253-257.
[26] Aas P. 2003. The threat of mild-spectrum chemical warfare agents. Prehosp Disast Med 18(4):306-312.

Scientific Networking and the Global Health Network Supercourse
F. Linkov and R. LaPorte (Eds.)
IOS Press, 2006
© 2006 IOS Press. All rights reserved.

Biosafety and Biosecurity in Georgia

Lela BAKANIDZE, P.G. IMNADZE, SH.A. TSANAVA, N.S. TSERTSVADZE
National Center for Disease Control and Medical Statistics of Georgia, 9 Asatiani Street, Tbilisi 0177, Georgia

Abstract. The legitimate functions of science, industry, medicine, and public health must be preserved, while at the same time addressing the risks they engender. In order to achieve this, there must be an understanding that the traditional boundaries between health and security need to be redrawn. The threat of infectious disease—whether naturally occurring or as a result of deliberate use —is real and growing, and the global community must find appropriate ways to counter this threat without crippling the very institutions that are critical to health security. Georgia, as part of former Soviet Union, used to share legislation including regulations on biosafety. Even though it's been more than decade since Georgia became independent, and while new Georgian laws have been adopted (parts of sanitary norms and regulations were renewed), many laws remain unchanged. The need for new Georgian regulations on biosafety/biosecurity is evident. The draft of new Georgian legislation on biosafety includes four documents: (1) Select Agents Rule; (2) Rules of Import, Export, Containment, Transfer and Handling of Cultures of Infectious Diseases Causative Agents (Bacteria, Viruses, Rikketsia, etc.), Protozoa, Mycoplasma and Genetic Materials, also Toxins and Poisons of Biological Origin; (3) Sanitary Norms For Labs Working with Especially Dangerous Pathogens; and (4) Guidelines for Safe Transportation of Infectious Substances and Diagnostic Materials. This new legislation package will be supported by all involved agencies like the Central Sanitary Inspection of the Ministry of Labor, Health and Social Affairs of Georgia, Ministry of State Security of Georgia, Ministry of Infrastructure of Georgia and Ministry of Interior of Georgia. Legislation on biosafety/biosecurity, like all the other legislation, rules, and regulations can be ideal, even applicable, but there must be readiness in the community to follow them. This requires creating mechanisms for their implementation.

Keywords. Biosafety, biosecurity, Georgia, pathology, public health

Introduction

The threat of infectious disease—whether naturally occurring or as a result of deliberate use—is real and growing. Expanding globalization and advancing biotechnology drive the threat. The global community must find appropriate ways to counter this threat without crippling the very institutions that are critical to health security. Recent natural outbreaks of highly infectious diseases have had devastating consequences for public health, the international economy, and international security. The consequences of an outbreak of infectious disease resulting from the biological weapon use would be at least as damaging as a naturally occurring infection, and even worse.

The risk of infectious disease resulting from an accidental release of a pathogen from a laboratory setting or use of biological weapon is real and growing. The rapid expansion of the biotechnology industry has resulted in the global proliferation of bilateral use of biological materials, technologies, and expertise. As a result, dangerous pathogens are more accessible to a wide range of those who are interested in proliferating biological weapon (terrorists being among them).

Many different methods are being used to address global risks associated with naturally occurring and accidentally or intentionally introduced infectious diseases. Most strategies—like increasing the effectiveness and availability of therapeutics, improving diagnostic capabilities, and developing decontamination and detection technologies—focus on enhancing national response to an outbreak of infectious disease after it had occurred.

It is difficult to identify mechanisms to counter biological threats, as no international framework exists, within which they can be applied (on the contrary to other threats, like the International Atomic Energy Agency oversees the Nuclear Non-proliferation Treaty, the Organization for the Prohibition of Chemical Weapons oversees Chemical Weapons Convention).

Nearly all current strategies for biological security are national in origin and focus on the response to disease outbreaks. These include increasing the effectiveness and availability of vaccines and other therapies (such as antibiotics), improving disease surveillance and diagnostics, and developing improved decontamination and detection technologies; and building public health capabilities.

The U.N. Security Council Resolution 1540, adopted on April 28, 2004, under Chapter VII of the U.N. Charter, was motivated by the concern that non-state actors "may acquire, develop, traffic in or use nuclear, chemical and biological weapon sand their means of delivery." Security Resolution 1540 recognises "the need to enhance coordination of efforts on national, sub-regional, regional, and international levels in order to strengthen a global response to this serious challenge and threat to international security."

Georgia, when part of the former Soviet Union (FSU), shared legislation including regulations on biosafety that dealt with especially dangerous pathogens, with other republics of the country. Biosafety rules in the FSU were very strict, and they did not leave gaps for misunderstanding; bioweapon facilities were well defined, they were components of Soviet offensive biological weapons program, and no other institutions were authorized to have any especially dangerous pathogens.

More than decade has passed since Georgia became independent, but not the only parts of the biosafety legislation has been adapted to fit the new situation. New Georgian laws were adopted, part of sanitary norms and regulations were renewed, but many parts remain unchanged. For example, even classification of pathogens by risk groups is absolutely contrary to the WHO classification. Today, when the threat of bioterrorism is more realistic, the need for new regulatory basis for biosafety has become evident. Furthermore, scientific investigations on especially dangerous pathogens that are very well funded by different donors, often do not take into account whether proposed recipients of their grants are eligible to carry out such investigations, and whether they have sufficient experience to carry such activities

In what was the Soviet Union, there were only a few biological weapons-related institutions in Georgia (compared to Russia and Kazakhstan) including Georgian Anti-Plague Station (at present National Center for Disease Control and Medical Statistics [NCDC] of Georgia); the Ministry of Labor, Health and Social Affairs of Georgia was the only institution that carried out surveillance on especially dangerous infections (plague, anthrax, tularemia) throughout the territory of Georgia; Biokombinat, in Tabakhmela (near Tbilisi – capital of Georgia), produced live vaccines for foot and mouth disease, etc., and; partially Eliava Institute of Bacteriophage, manufactured different vaccines (rabies, anthrax, etc). These institutions worked under strict regulations. After the collapse of Soviet Union there was a significant outflow of military microbiologists to other institutions, often to private companies, and this is a source of concern.

1. Problem Statement and Objectives

As we have already mentioned, Georgia, as the part of Soviet Union, shared legislation with the other republics including regulations on biosafety/biosecurity. Now, more than a decade has passed since Georgia became independent, and the biosafety/ biosecurity legislation has not been worked out or approved. New Georgian laws (like the Georgia law "on Health Care," the Georgia law "Georgian Sanitary Code," etc.) were adopted with parts of sanitary norms and regulations renewed, but the majority of the laws remain unchanged.

The protection of dangerous pathogens and toxins from theft and sabotage at the facilities where they are used and stored provides the first line of defence against both biological weapons proliferation and bioterrorism by making it more difficult for proliferators to acquire dangerous biological materials.

The term "laboratory biosecurity" is newly introduced, and it is often confused with the more widely recognised term "laboratory biosafety" (in Georgian, both terms are expressed by one word). Laboratory biosecurity and laboratory biosafety—are critical to the operation of the modern bioscience laboratory—often overlap and should complement each other, though they have different objectives:

Laboratory biosafety reduces biological risk, aims to reduce or eliminate exposure to laboratory workers or other persons, and exposure to the outside environment to potentially hazardous agents involved in bioscience or biomedical research. Laboratory biosafety is achieved by implementing various degrees of laboratory "containment," or safe methods of managing infectious materials in a laboratory setting.

Laboratory biosecurity aims to protect pathogens, toxins, and security-related information from theft and sabotage. Laboratory biosecurity is achieved by instituting a culture of responsibility among those who handle, use, and transport dangerous pathogens and toxins, and by implementing various security measures that restrict access to these materials to authorized individuals. Although it is impossible to say how likely it is that there will be a bioterrorist attack or the bio-crime in Georgia, the likelihood is growing.

The need for new Georgian regulations on biosafety /Biosecurity is evident, as Georgia continues to abide by old Soviet regulations in handling dangerous pathogens: (1)The Decree of the Ministry Of Health of USSR "Concerning Rules of Registration, Containment, Handling and Transfer of Cultures of Pathogenic Bacteria, Viruses, Rikketsia, Fungi, Protozoa and others, also Bacterial Toxins and Poisons of Biological Origin" approved by the Ministry of Health of USSR, 18.05.79. and (2) the "Instruction on Regime of Control of Epidemics while Working with Materials Infected or Suspected to be Infected with Causative Agents of Infectious Diseases of I-II Groups" approved by the Ministry of Health of USSR 29.06.1978.

2. New Georgian Legislation on Biosafety/Biosecurity

NCDC Georgia was established on the basis of the Georgian Anti-Plague Station of the Ministry of Health of USSR. NCDC Georgia owns the National Collection of Especially Dangerous Pathogens. Among its other duties, the NCDC Georgia, according to the Decree of the President of Georgia No. 55, February 21, 2003, in the statute "Participation in drafting normative and methodological documents on surveillance, disease control and prevention, biosafety/biosecurity," the Department of Biosafety and Threat Reduction, together with experts from other departments of the Ministry of Labor, Health and Social Affairs of Georgia, prepared the draft of the Georgian biosafety /biosecurity legislation. It must be mentioned that during the work on legislation, we had great help from U.S. Department of Defense Threat Reduction Agency (DTRA), the U.S. Department of Health and Human Services (DHHS), and the U.S. Department of State.

While working on the new legislation draft, we had found that improving biosecurity in the country will be more difficult than previously thought. It required more legislative changes, like changes in the "Criminal Code of Georgia" and "Georgian Administrative Code" where the penalties for bio-crimes are insignificant, especially when compared to the bonuses that criminals may receive from those aiming to obtain especially dangerous pathogens. Involvement of law enforcement agencies must be more active.

New Georgian legislation on biosafety/biosecurity is based on WHO and American regulations on biosafety, particularly the "Laboratory Biosafety Manual," World Health Organization (WHO/CDS/CSR/ LYO/2003.4 and "Biosafety in Microbiological & Biomedical Laboratories, US DHHS, CDC Atlanta & NIH, Fourth Edition, May, 1999).

We created the new legislation to be in compliance with Georgian national legislation, particularly:

1. Law of Georgia "on Health Care" (10.12.1997) Cl.70 states: "Providing safety for public health environment is the responsibility of the State. The Ministry of Health of Georgia elaborates, approves sanitary-hygiene regulations and norms and controls their observance"; Cl.72: "Observance of sanitary-hygiene, sanitary-control regulations and measurements elaborated for avoiding negative effects on the environment or other factors on public health that are approved, is obligatory for any physical, or legal body in spite of its proprietary, organizational or legal form or departmental subordination"; Cl. 77: "Import, export, containment, transfer and work with infectious diseases causative agents is allowed only under permission of the Ministry of Health of Georgia".
2. Law of Georgia "Georgian Sanitary Code"(08.05.2003) regulates legal relations concerned with maintaining safety for the human health environment draws ways of carrying out state control on implementation of sanitary norms and preventive sanitary-hygiene and sanitary-control measurements.
3. Law of Georgia "on Export Control of Armament, Military Techniques and Products of Bilateral Purpose" (28.04.1998) among products undergoing export control in Cl.4d: "Causative agents of diseases, their genetically modified forms and fragments of genetic materials that can be used for production of bacteriological (biological) and toxic weapons according to the list of international regimes of nonproliferation."

It was decided, that Georgian legislation on biosafety will be a package consisting of four documents including:

1. Select Agents Rule;
2. Rules of Import, Export, Containment, Transfer and Handling of Cultures of Infectious Diseases Causative Agents (Bacteria, Viruses, Rikketsia, etc.), Protozoa, Mycoplasma and Genetic Materials, also Toxins and Poisons of Biological Origin;
3. Sanitary Norms For Labs Working with Especially Dangerous Pathogens;
4. Guidelines for Safe Transportation of Infectious Substances and Diagnostic Materials.

Select Agents Rule is the main tool for regulating the work of laboratories with select agents—especially dangerous human and animal pathogens (a list of overlapping select agents is now being coordinated with veterinary services). The basis for this tool relied on the U.S. Federal Register Part IV, DHHS "Possession, Use, and Transfer of Select Agents and Toxins; Interim Final Rule." It draws mechanisms of registration, security risk assessments, safety, security, emergency response, transfers, record keeping, inspections, duties of responsible official, training, notifications for theft, loss or release, administrative review, criminal penalties (here sufficient chapters from the Administrative Code of Georgia and the Criminal Code of Georgia were used), submissions and forms, applicability and related requirements.

As a basis for "Rules of Import, Export, Containment, Transfer and Handling of Cultures of Infectious Diseases Causative Agents (Bacteria, Viruses, Rikketsia, etc.), Protozoa, Mycoplasma and Genetic Materials, also Toxins and Poisons of Biological Origin," "Sanitary Norms For Labs Working with Especially Dangerous Pathogens," and "Guidelines for Safe Transportation of Infectious Substances and Diagnostic Materials" used WHO and U.S Centers for Disease Control and Prevention as source information for regulations.

This new legislation package will be discussed and developed with the assistance of all involved agencies including the Central Sanitary Inspection of the Ministry of Labor, Health and Social Affairs of Georgia, the Ministry of State Security of Georgia, the Ministry of Infrastructure of Georgia, and the Ministry of Interior of Georgia.

3. Summary and Conclusions

Pathogenic microorganisms, naturally occurring or man-made, present a significant threat to global health security. There is no clear consensus on how to globally limit the distribution or availability of pathogenic microorganisms. Development of national and international norms for the security of these organisms are complicated by the fact that they appear naturally and globally, can self-replicate, are difficult to detect at distance, and can be used for harm and good. Development of security standards should be broad-based and take into consideration the needs of health, science, economics, and national and international laws.

New legislation on biosafety, like all the other legislation, rules and regulations, can be ideal, even applicable, but there must be readiness in the community to follow them. Furthermore, mechanisms for implementation must be created.

References

[1] Law of Georgia. On Health Care. (10.12.1997).
[2] Law of Georgia. Georgian Sanitary Code. (08.05.2003).
[3] Law of Georgia. 1998 April 4. On Export Control of Armament, Military Techniques and Products of Bilateral Purpose.
[4] Ostfield M. 2004. Bioterrorism as a foreign policy issue. SAIS Review 24(1):131-146.
[5] Heymann DL, Aylward RB, Wolff C. 2004. Dangerous pathogens in the laboratory: from smallpox to today's SARS setbacks and tomorrow's polio-free world. The Lancet (363) 16566 – 68p.
[6] Steinbruner D, Harris ED. 2003 Spring. Controlling dangerous pathogens. Issues in Science and Technology; 47-54 p.
[7] Biotechnology research in an age of terrorism: confronting the dual use dilemma, National Research Council of the National Academies. 2003. National Academy Press.
[8] Tucker JB. 2003 November. Biosecurity: limiting terrorist access to deadly pathogens. US Institute of Peace: Peaceworks; 52.
[9] Salerno RM. 2005. The US Select agent rule and an international opportunity to develop laboratory biosecurity guidelines proceedings of the international seminar on nuclear war and planetary emergencies, 32nd Session, Erice. World Scientific Press; 221-228 p.
[10] Bakanidze LG, Imnadze PG, Tsanava ShA, Tsertsvadze NS. 2005. New Georgian Legislation on Biosafety. Proceedings of the International seminar on Nuclear war and Planetary Emergencies, 32nd Session, Erice. World Scientific Press; 229 – 231 p.

Scientific Networking and the Global Health Network Supercourse
F. Linkov and R. LaPorte (Eds.)
IOS Press, 2006
© 2006 IOS Press. All rights reserved.

Safety Issues of Electric Power Production in the Russian Federation: Brief Review and Analysis of Accidents and Emergencies

Irina PROKOPYEVA [a], Andrey TRUFANOV [b]
[a] New-Zima Heat-Electric Generating Station of Irkutskenergo Corp, Sayansk, Irkutsk Region, RF
[b] Irkutsk State Technical University, Irkutsk, RF

Abstract. A comparison of the reliability of the electric energy system in Russia with pertinent U.S. experiences is discussed. Unique digits of accidents and emergencies at the third Russian electricity production corporation, Irkutskenergo, in 2004 are reported. Networking and Information Security approaches are argued as helpful in solving problems of disaster prevention in the industry.

Keywords. Russia, Electric energy system, accidents, networking, prevention, information security

Introduction

National security is linked in innumerable ways to accessible, secure, and preferably cheap energy upon which industries and armed forces rely. Energy supply in the form of electricity operates most modern technologies and continuity is required to prevent disruption in today's society. Simply put, electricity is an essential need of modern life. One need only consider the consequences of a relatively short power outage—factories, offices, and stores close down; many ICT devices go dead; traffic comes to a crawl or standstill; food spoil in freezers and refrigerators; homes are dark—to perceive our dependence on electric power. Some critical industries, such as hospitals, rely almost exclusively on electricity, as it is the lifeblood of a hospital. Naturally, patients' lives depend upon sophisticated medical equipment and highly skilled medical professionals, but people and equipment depend, in turn, on a hospital's electrical power system. Without power, lighting would be inadequate, temperature control would be difficult, and most critically, certain types of life support and diagnostic equipment would be impossible to use.

On the whole, energy systems could be referred to as national critical infrastructure. The National Strategy for Homeland Security in the United States categorizes critical infrastructures into the following sectors [1]: Agriculture, Food, Water, Public Health, Emergency Services, Government, Defense Industrial Base, Information and Telecommunications, Energy, Transportation, Banking and Finance, Chemical Industry and Hazardous Materials, Postal and Shipping.

The critical infrastructures typically interconnect and, therefore, depend upon the continued availability and operation of other systems and functions. For example, E-commerce depends upon electricity as well as information and communications.

Thus, accidents in electric industry result in the following consequences:

- Outage of consumers;
- Influence on sensitive institutions (medical, sanitary, air traffic control, law-enforcement etc.);
- Transportation problems ;
- Outage of ICT devices;
- Outage of food storage equipment;
- Influence on system failures of functioning of dangerous industrial units;
- Emergency protection and action of the personnel on nuclear energy units;
- Environmental damage.

Accidents and emergencies, in any critical industry, can be caused by natural disasters, technical failures, personnel errors, or terrorist attacks.

In general, the electrical utility infrastructure, in any country, has a system of generation, transmission, distribution, communication, and other facilities that are physically connected and operated as a single unit under one control. The flow of electricity with the system is maintained and controlled by dispatch centers. It is the responsibility of the dispatch center to match the supply of electricity with the demand for it. As a rule, in order to carry out its responsibilities, the dispatch center is authorized to buy and sell electricity based on system requirements. Electrical power systems have their specificity: electricity cannot be stored; it must be generated, transmitted to the consumer, and consumed instantaneously. Electric utility companies worldwide are treated as monopolies because duplicate facilities, particularly transmission and distribution lines, would be inefficient. An accident at one unit impacts on the entire electrical system and might cause a crisis in vast regions. Furthermore, events happen faster than experts are able to react to them.

1. Reorganization of the electric power system in Russia

Today, the electric power industry of Russia is in a stage of restructuring, which notably relies on the management of electrical power systems. The common philosophy in management can be generalized by the following phrase: "from technical operators to organizers." Privatization and market competition have resulted in a radical change of a control system. While technical decisions are still carried out on the engineering level, financial issues demand revisions of many traditional approaches.

In July, 2001, the government of the Russian Federation reached Decision 526: About reforming electric power industry of the Russian Federation, which divided the uniform electric power complex into numerous commercial structures for generating, transmission, distribution, and selling electricity. The decision mainly reflected interests of the management of RAO UES (the United Energy System of Russia) Corp. In 2003, the law About Electric Power Industry implemented the liquidation of regional power joint-stock companies—associations of federal power generating stations of the wholesale market in incorporated generating companies, and also the break up of the centralized system of dispatching the management of the country's electric power complex.

2. U.S. Experience in the Electric Power Industry

The U.S. power industry is experienced in dealing with the conditions of the market:

- There are five functioning independent system operators (ISOs) in the U.S.
- 16, 756 generating companies in the U.S. have a total capacity of 1 million MWts.
- Transmission of electric power is carried out through interconnected networks.

The U.S.'s bulk power system is structured into three major networks (power grids), which also include smaller groupings or power pools. The major networks consist of extra-high-voltage connections among individual utilities designed to permit the transfer of electrical energy from one part of the network to another. The three networks are:

- Eastern Interconnected System;
- Western Interconnected System, and;
- Texas Interconnected System.

Virtually all U.S. utilities are interconnected with at least one other utility by these three major grids. The bulk power system makes it possible for utilities to engage in wholesale (for resale) electric power trade. However, wholesale trading does not lead automatically to greater reliability. So, on August 14, 2003, a series of power surges over a 12-second period triggered a cascade of shutdowns at more than 100 generating plants throughout eight states and Ontario, Canada. The result was the biggest blackout in North American history. 61,800 megawatts of power were lost to over 50 million people.

3. Reliability of the Russian Power Supply System

For Russia, with its severe climate and extremely high level of energy consumption, the electric power industry is especially important. Confidently, Russian market-partisan experts ruled out the possibility of a power interruption within the country: in the U.S. there is no integrated power system providing operative redirection of the electric flow from one region to another. Moreover, the American system has no uniform dispatching center similar to the Russian system operator. In Former Soviet Union (FSU) countries, power systems work in parallel: In 1998, NIS members signed an agreement and supported the work that reduces risk of outages up to a minimum.

Contrary to above-stated arguments Russia has experienced many large-scale failures in electric supply in various regions:

- May 2005, approximately 1,000 residential blocks in the south, southeast and southwest, and downtown Moscow did not have electricity, 113 kindergartens and 106 schools, one maternity home, one tuberculosis sanatorium, and other establishments (a total of 219 educational establishments and three healthcare facilities) had no electricity.
- June 2005, there were electricity outages in Dubna, Moscow region, due to a hurricane that damaged power transmission lines. In Blagovestshensk, Amur region, an energy energetic step generated the outage of a central transmission substation, and caused cascade outage of smaller substations; regional hospitals had no electricity for one hour as its emergency power supply equipment failed.

- July 2005, an electricity outage Vladivostok resulted because of a severe traffic accident—a crane truck damaged a high voltage power transmission line.

Data on accidents and emergencies, as a rule, are inaccessible for analysis by independent and nonpartisan experts as the result of the usual corporate policy of information restrictions. There is no data on accidents on the map of the Web site of RAO UES Corp. [2], nor in the report of the federal network company of Uniform power system [3]. The Russian Search Engine Yandex [4] proposed four sites with no references to concrete figures and data, even when it is required for "Accidents and emergencies in electric power industry."

4. Accidents and Emergencies at the Electric Power Production Corporation Irkutskenergo in 2004

The Corporation, the third largest power generating company in the Russian Federation, and the first one in the Asian part of the country, distributes electricity through its own high-voltage grid to 325,000 customers in the Irkutsk area. The Generator Nameplate Capacity (MW) of the Irkutskenergo Corp. is: 12881, 6 MW. This corporation has experienced its share of failures and incidents.

In 2004, the units of Irkutskenergo produced 53, 717 billion kWh; united three hydroelectric power stations, 12 thermal power stations, five electric grids, three heat grids, one trading company, one renovation company, Irkutskenergoremont, and a civil engineering company, Irkutskenergostroy; employed 12,000. The cost of electric energy for 2004 was 0.7 cent/kWh. The tariff for the electric power for urban population made 1.1 cent/kWh.

In 2004, one unit, New-Zima Heat-Electric Generating Station of Irkutskenergo Corp., delivered heat for its main consumer—Sayanskchemplast—for the population of Sayansk-city, the enterprises of agriculture, and simultaneously generated electric power, which flowed into the system of Irkutskenergo. Its Generator Nameplate Capacity was: electric, 240MW; heat, 3.631 G BTUs/h. About 600 persons work for the station.

Technological Infringements

The total number of technological infringements including failures and incidents at Irkutskenergo Corp., in 2004, are presented in the following (data for 2003 are given in brackets): Incidents 202 (266), Principal accidents (1), Total short-delivery of electric energy, 2140.6 (818.41) MWh., Total short-delivery of heat energy, 2.7 (0.67) GBtu, Economic damage on all infringements, -4,368 (4,170) M rubles.

a) Technological infringements:

- 1 (1) Infringement of treaty obligations on delivery of energy
- 1 (0) Damage of technological buildings, constructions,
- 1 (0) Full dump of electric or thermal loading of unit
- 103 (99) Damage of the equipment of unit
- 51(73) Damage of the equipment of electric grid of kV and higher
- 3 (1) Damage of the equipment of heating grids
- 16(11) Error reactions of protection automatics

- 38 (84) Outage of the equipment because of inadmissible deviations of technological parameters
- 0(1) Infringement of functioning of dispatching and technological control tools for the term of more than 1 hour (loss of connection of dispatching service with controlled object)

b) Infringement of conditions of safe operation of a dangerous industrial target:

- 66 (44) The boiler equipment
- 3 (4) The turbine equipment
- 2(0) The auxiliary thermal-mechanical equipment
- 9(9) The electrotechnical equipment
- 1(0) natural disaster impact

c) Infringements of organizing rules and codes

- 11 (16) Inaccurate or Unauthorized actions of operators
- 6 (9) Inaccurate or Unauthorized actions of the personnel of services
- 15 (8) Inaccurate or Unauthorized actions of the involved personnel
- 10(13 Inaccurate or Unauthorized actions of repair personnel
- 2 (2) Inaccurate or Unauthorized actions of chief personnel
- 2 (2) Unsatisfactory quality of instructions,
- 7 (10) Default of volumes and terms of maintenance service
- 25 (34) third party (organizations and persons) impacts,
- 16 (57) natural disasters impacts

These data are delivered to safety departments of units of Irkutskenergo Corp.

5. Information Security Approaches for the Decision of Problems of Power Industry

Naturally, technical solutions and advanced load control techniques to prevent accidents are welcomed in the industry. But, it becomes clear that the most basic issue is not a technological one; it is not a gizmos problem, it is a people problem.

Instruments to prevent accidents in the electric power production industry can be classified the same way as in InfoSec: Ethical, Legislative, Organizational, Physical (technical facilities and equipment), Technical (hardware and software), Math.
InfoSec claims that renovation of a system makes that vulnerable—it just provides warnings about the reorganization of the electric power system in Russia.

If we enumerate only some points and principles in InfoSec, one finds them of sufficient value for accident prevention in the electrical industry: Security Policy, Analysis of threats, Interdisciplinary Approach and Networking, Personnel Security;

The most powerful tool cited by InfoSec is networking. Networking could be applied to prevent accidents in the electric power production industry. Networking involves not only technical peer-to-peer connections inside the corporation but effective, reliable, and productive information exchanges among diverse disciplines (interdisciplinary Networking), diverse management level (inter-level Networking) and throughout the world (international Networking).

The following positive aspects can be listed for implementing networking prevention into the electric energy industry of Russia: Restructuring the industry,

Modern ICT, current experience of the U.S. electricity market, and educated personnel. On the other hand, there some negative aspects:

- anti-networking—separation created intentionally;
- execution of natural networking foundation—the changing electric power industry ;
- new management scheme—based on family-friend interest structure.

In the Russian industrial environment it is very problematic to make theoretical networking schemes work because of the set of threats and obstacles. The problems are social, not technological.

6. Summary

During the era of the Soviet Union, the main electric power industry feature was generating electricity, characterized by centrally controlled energy delivering structures. That inexorable scheme of control was technically reliable, and a significant part of this framework remains intact today.

Secondly, the introduction of market mechanisms and the separation of the industry into three components: generation, distribution and trading, has made the traditional methods incapable of providing former reliability in the industry. It is impossible to freeze the re-structuring of the former control system; therefore, renovation requires effective means of reliability of operations.

Third, the point is not in the type of the market, but whether it is controlled by society (through the government); networking is necessary for management to prevent accidents and emergencies.

Finally, critical infrastructures should be reliably protected. Protection should be provided at all levels—government, corporate and personal. Therefore, it seems reasonable to support networking as interdisciplinary, inter-level, and international, allowing the use of experience of another industry, and of other levels and other nations.

References

[1] The National Strategy for the Physical Protection of Critical Infrastructures and Key Assets. 2003. February.

[2] RAO UES of Russia Website. 2006. [Accessed March 30.] Available from: http://www.rao-ees.ru/ru/internet/map.cgi

[3] FSK EES Web site. 2006. [Accessed March 30.] Available from: http://www.fsk-ees.ru/ru/

[4] Yandex Search Engine Web site. 2006. [Accessed March 30.] Available from: http://www.yandex.ru/

Scientific Networking and the Global Health Network Supercourse
F. Linkov and R. LaPorte (Eds.)
IOS Press, 2006
© 2006 IOS Press. All rights reserved.

Chernobyl Disaster and Experience of Population Protection from Nuclear Accident

B.A. LEDOSHCHUK[a], N.A. GUDZENKO[b]
[a]*Department of Public Health,, National Medical University*
6,Yakubobskiy Str., Kiev, Ukraine
[b]*Research Center for Radiation Medicine*
53, Melnikova Str., Kiev, 04050, Ukraine

Abstract. The paper presents main functioning results, feasibilities and limitations of the State Registry of Ukraine for the Chernobyl NPP accident survivors. The structure of the individual records, completeness, and quality of the registered data allow assessing of the demographical and health status changes trends in victims forming the cohorts for profound research of the health outcomes. Registered data on doses received and diseases diagnosed in victims are of uncertain quality and require additional verification.

Keywords. Accidents, nuclear disasters, public health

Introduction

Minimizing issues of hazardous factor isolation and health detriment in immediate and remote terms are a top priority when it comes to man-made, ecological, and other humanitarian disasters. Establishment of population registries provides monitoring of the population involved in a disaster, even if there are hundreds and thousands of survivors. Registries of the A-bomb survivors in Japan and the registry launched in the former Soviet Union (FSU) following the events of the Chernobyl disaster are prime examples of the most successful large-scale registries. A registry of survivors provides the estimation of current detriment to population health after the accident, the forthcoming situation, and simulating and establishing an informational basis for decision making on a population level. Launching the registry of survivors enables accounting for the social interventions and health monitoring activities, as well as for organizing the required healthcare maintenance activities. Moreover, the survivors' registry data are a unique informational source for current, retrospective, and prospective scientific estimates of accident or disaster effects.

1. Objectives

The paper has the following objectives: to highlight the role of the State Chernobyl Registry of Ukraine (SChR) as a part of the health monitoring and protection system for survivors after the Chernobyl Catastrophe and; assess the feasibilities and limitations of the State Chernobyl Registry and its applicability for conducting epidemiological studies in Chernobyl disaster clean-up workers

2. Historical background

The largest wide-scale, man-made disaster of the second half of 20th century occurred in Ukraine at the Chernobyl nuclear power plant in April 1986. The Chernobyl catastrophe required immediate decision making steps to limit and minimize detrimental effects on the environmental and the human population.

Outstanding prophylactic or preventive measures were performed in many regions of the Ukraine, Russia, and Belarus to minimize population irradiation including the evacuation of the civil population and medical monitoring of sufferers. The total number of the resettled persons was more than 150,000.

Chernobyl catastrophe Victims comprise four main groups

- Group 1: persons involved in the clean-up operations at the Chernobyl Nuclear Plant (liquidators)
- Group 2: Persons evacuated from the exclusion zone in 1986 (evacuees)
- Group 3: Persons residing in the territories monitored (relocation zone) or residing there immediately after the accident (residents)
- Group 4. Children born to parents in Groups 1-3 (offspring)

As logistical arrangements of launching the registry of survivors took several months to establish after the catastrophe, some extent of primary data had been irreversibly lost.

The SChR is the legal successor to the Ukrainian section of the All-Union Distributed Clinical-Dosimetry Register, founded in the FSU. Health state monitoring among survivors is the main SChR objective, aiming to improve decision making on medical provisions of survivors and their social security benefits.

The following informational components were accounted for in the registry design:

- Data enabling identifying the survivor, tracing him or her under temporary or unvarying migration, linkage conduction with other population registries (e.g., social, nosological, occupational ones); those data include full name, place of residence, gender, date of birth, date of registration, registration group, number of the victim certificate
- Data on accident factor(s) impact (place of stay in the contaminated area, period of stay in the contaminated area, dose of the external irradiation, thyroid dose);
- Information on possible outcomes of factor impact (diseases or health disorders through generally accepted classification systems such as ICD-9 and ICD-10)

The SChR is a unique informational system providing survivor identification to trace individual health outcomes. Due to its three-level structure (regional, oblast, and state levels), SChR links the primary level of health care (doctors in charge of the case, district/regional hospitals), and central data bank, which summarizes the data and ensures information safety and security.

The SChR has been opened for inclusion of survivors. Due to the modified inclusion criteria in the recent years, the amount of registered persons has increased. As of January 1, 2004, the number of registered persons was over 1.5 million, including over 1 million residents of contaminated territories, and 279,000 clean-up workers (further-liquidators). By the middle of 2005, the number of registered persons grew to more than 2 million. Therefore, the necessity of monitoring the scale of complexity emerges both with data verification degree, technical, expert and financial maintenance.

3. Completeness and Quality of the Chernobyl Registry Data on Entries Related to Male Clean-up Workers

Clean-up workers attract special attention as the most irradiated group after the Chernobyl accident; these are persons of the most active social and productive age. Most of them are males (about 90 %) who were irradiated with a low or moderate dose of external irradiation (up to 0.2 Gy), but some of them—with much higher doses (up to 1 Gy and even more).

The completeness of individual registration data elements on "liquidators" in the SChR is in a range of 82.5–100.0%. It allows for identifying the liquidator, surveying his status and migration, tracing his current location, and establishing contact with him, if necessary. In addition, it helps to form cohorts for the detailed medical follow up and scientific studies. Based on these data, one may assess demographic structure of liquidators (see Figure 1).

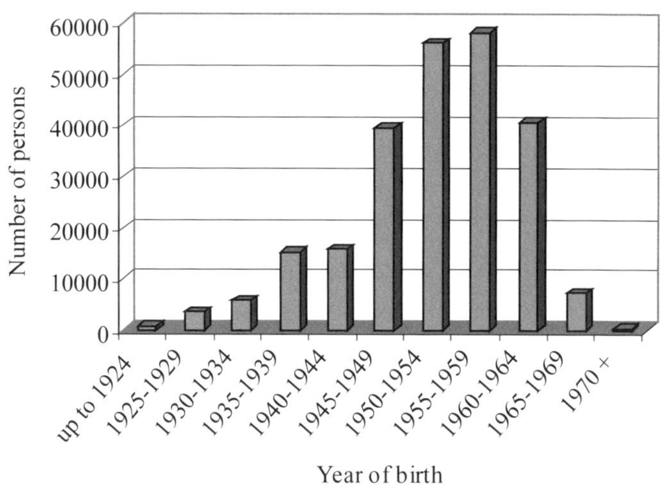

Figure 1. Age structure of the male liquidators registered at SChR

The majority of liquidators who participated in clean-up activities were young, with 80 percent were born between 1945-1964.

Elements of information on those who stayed in the contaminated area and doses of irradiation received are of different completeness. Twelve percent of individual records do not have the data on the period of clean-up activity. According to available data: 53.5 % of liquidators participated clean-up work in 1986; 19.9 % in 1987; 8.4 % - в 1988 г.; 5.3 % in 1989; 1.6 % - в 1990 (see Figure 2).

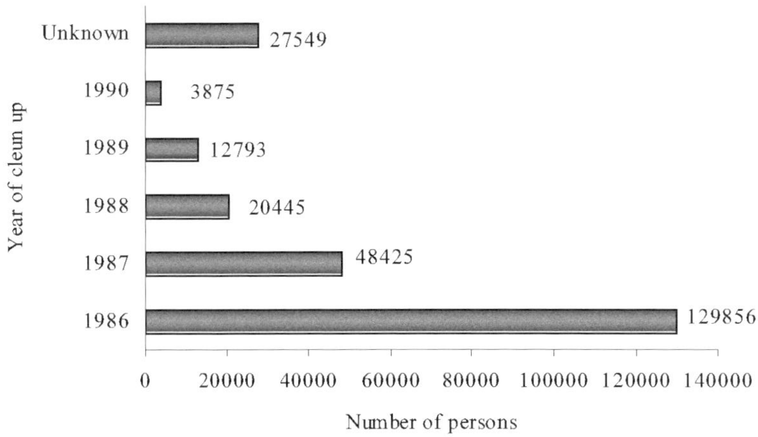

Figure 2. Distribution of liquidators by year of clean up

Data on doses received are available in 42 percent of individual records with known year of clean up including 29 percent of liquidators who participated in clean up work in 1986. Tables 1and 2 show the presence of dose records and dose distribution in liquidators by year of their activities.

Table 1. Presence of the dose records in liquidators at the SChR

Year of clean up	Doses available		Doses not available		At all	
	Abs	%	abs	%	abs	%
1986	37331	28,6	93031	71,4	130362	100,0
1987	28475	58,8	19949	41,2	48424	100,0
1988	12969	63,4	7471	36,6	20440	100,0
1989	8857	69,2	3935	30,8	12792	100,0
1990	2384	61,5	1491	38,5	3875	100,0
1986-1990	90016	41,7	125877	58,3	215893	100,0
Unknown	160	0,6	26893	99,4	27053	100,0
Total	90176	37,1	152770	62,9	242946	100,0

Incompleteness and uncertain precision of the registered doses do not allow the use these values for the dose dependent risks assessment.

Table 2. Available dose distribution by year of clean up (%)

Year of clean up	Dose value (cGy)						
	< 5	5,0 - 9,9	10,0-14,9	15,0-19,9	20,0-24,9	25,0-49,9	50,0 +
1986	7,8	8,6	10,5	19,3	33,9	19,3	0,5
1987	9,7	43,5	26,8	6,1	11,9	1,8	0,2
1988	41,1	52,2	5,4	0,5	0,4	0,3	0,2
1989	30,8	67,7	0,6	0,2	0,2	0,3	0,1
1990	39,8	59,2	0,4	0,1	0,1	0,1	0,3
Unknown	25,6	20,6	8,1	15,6	22,5	5,0	2,5
Total	16,3	33,0	13,7	10,0	17,9	8,7	0,4

Annual health check-up data completeness for registered liquidators is on average 90 percent, but the quality of input medical data is uncertain and requires further diagnostic review. This makes it impossible to use this information directly for the detailed epidemiological studies of Chernobyl Disaster health outcomes.

The incompleteness of health check-up data entered into the SChR database reflects the real situation in the health state of survivors. Excessive registration of diseases is present including miscoding, registration of cases that are unconfirmed at later stages, diagnostic errors, and computer data input errors.

Based on the review of the results on the diseases of special interest in terms of a possible link with irradiation (leukemia), more then half of registered cases were false positives. Registration delays may vary from one to six years.

4. Conclusions

The SChR for persons who survived the Chernobyl disaster is a sufficient tool for the registration and follow up of Chernobyl disaster victims. It is the first-rate population registry of the country covering the entire territory and including, on the individual level, the extent of information capable to provide informational basis for the assessment of demographical and health status changes trends.

As an informational database, SChR provides opportunities for the development of cohort studies, such as detailed research on the Chernobyl disaster effects including tracing of the study subjects and establishing contact with them.

The use of the radiation doses values officially registered in the SChR for dose dependent diseases risk estimation is limited by incompleteness and uncertain precision. Included in the study cohort only the subjects having dose records can substantially decrease the study power, and could lead to bias in disease risk estimates under investigation. Such analysis requires elaboration of universal (suitable for both living and deceased subjects) method of radiation dose reconstruction as recently proposed in the questionnaire based method.

The SChR can be a supplementary informational source for cases of diseases of interest due to the over recording and delay in case registration within the entire functioning period. The sufficient way to monitor the diseases in Chernobyl victims is the combination of the data from specialized registries of Ukraine (SChR) with obligatory, independent verification of diagnoses.

References

[1] Concept of the national registry of Ukraine for persons survived after the Chernobyl disaster MH of Ukraine. 1992. Kiev.. (In Ukrainian).

[2] Gulak LO. 2001. Methodological aspects of the wide-scale cancer-registry establishing Oncology: Suppl. to J. Exp. Oncol (3)2-3:198-202p.

[3] Howe GR, Zablotska LB, Fix JJ, Egel J, Buchanan J. 2004 November. Analysis of the mortality experience amongst U.S. nuclear power industry workers after chronic low-dose exposure to ionizing radiation. Radiat Res 162(5):517-26p.

[4] Ivanov V, Tsyb A, Ivanov S, Pokrovsky V. 2004. Medical radiological consequences of the Chernobyl catastrophe in Russia: estimation of radiation risks. St.Peterburg: Nauka.

[5] Kryuchkov V, Anspaugh L, Bouville A, Bakhanova E, Cardis E, Chumak V, Golovanov A, Howe G, Luckyannov N, Voileque A. In press. RADRUE-method for dose reconstruction for Chernobyl liquidators in questionnaire-based studies. Radiat.Prot.Dosm..

[6] Ledoshchuk BA. 1991.National distribution registry of individuals exposed to radiation effects due to the Chernobyl AES accident. Vestn Akad Med Nauk USSR. (8):9-11. (In Russian.)
[7] Ledoshchuk BA, Trotsuk NK, Gudzenko NA, Bebeshko VG, Belyayev Yu, Dyagil N. 2003. I.S. Modern methodology on data quality control for the state registry of suffered resulting from the Chernobyl catastrophe.Information technologies in Public Health and practical medicine. ProcScientific papers. Proceedings of the conference, 21-23, May, 2003, Kiev: 47 – 51. (In Russian.)
[8] Mabuchi K, Soda M, Ron E, Tokunaga M, Ochikubo S, Sugimoto S, Ikeda T, Terasaki M, Preston DL, Thompson DE. 1994 February. Cancer incidence in atomic bomb survivors. Part I: Use of the tumor registries in Hiroshima and Nagasaki for incidence studies. Radiat Res 137(2 Suppl):S1-16 p.
[9] Romanenko AYe, Beebe GW, Bebeshko VG, Howe GR, Thomas TL, Ledoschuk BA, Kortushin GI, Gudzenko NA, Burch JD, Chumak VV, Bouville A, Pilinskaya MA, Klymenko VI, Dyagil IS, Finch SC, Reiss RF, Bazyka DA, Masnyk IG, Tirmarche M. 2003. Leukemia: results of a two year feasibility study. In: Health effects of Chornobyl Accident: Monograph in 4 parts/Ed. A.Vosianov, V.Bebeshko, D.Bazyka – Kiev: DIA.38-53p.
[10] Shilnikova NS, Preston DL, Ron E, Gilbert ES, Vassilenko EK, Romanov SA, Kuznetsova IS, Sokolnikov ME, Okatenko PV, Kreslov VV, Koshurnikova NA. 2003 June. Cancer mortality risk among workers at the Mayak nuclear complex. Radiat Res 159(6):787-98p.
[11] Shin H, Ramsay T, Krewski D, Zielinski JM. 2004 December 8. The effect of censoring on cancer risk estimates based on the Canadian National Dose Registry of occupational radiation exposure. J Expo Anal Environ Epidemiol [E-pub ahead of print.]
[12] Tsyb AF, Ivanov VK, Airanepetov SA, et al. 1991. Radiation-epidemiological analysis of the national register of persons exposed to radiation as a result of the Chernobyl accident. Vestn.Akad.Med.Nauk SSSR 11:32-36p. (In Russian.)
[13] UNSCEAR 2000 Report to the General Assembly. Annex J. Exposures end effects of the Chernobyl Accident Int. Journal of Radiation Medicine; 2000 2-4 (6-8).
[14] Winkelmann RA, Okeanov A, Gulak L, et al. 1998. Cancer registration techniques in the NIS of the former Soviet Union. IARC Technical Report No 35.

Scientific Networking and the Global Health Network Supercourse
F. Linkov and R. LaPorte (Eds.)
IOS Press, 2006
© 2006 IOS Press. All rights reserved.

Chernobyl Disaster Update: Networking Examples

B.A. LEDOSHCHUK[a]
[a]*Department of Public Health,, National Medical University*
6,Yakubobskiy Str., Kiev, Ukraine

Abstract. Nuclear accidents or disasters are quite different from other natural or man-made events because of their detrimental radiation impact and delay in the occurrence of conditions. One of the largest nuclear accidents took place in Chernobyl, Ukraine, in April 1986. This article describes lessons learned from this accident. The author also describes his collaboration with the Global Health Network Supercourse project and his experiences with sharing disaster data via the Internet.

Keywords. Disasters, nuclear accidents, Chernobyl, public health, Supercourse

Introduction

Nuclear accidents or disasters are quite different from other natural or man-made events because of their detrimental radiation impact and delay in the occurrence of conditions by several or dozens of years. The history of radiation accidents and disasters is not much older than 100 years. However, these accidents are directly linked to the technical progress, diagnostics, and management perfection in medicine, the nuclear industry, and in the arms race. The nuclear disaster that resulted from dropping the A-bomb on the Japanese cities, Hiroshima and Nagasaki, in 1945 led not only to public fear of nuclear attack, but also to anxiety regarding the aftermath of probable radiation.

No clear sign of physical injury within a few hours following exposure and poor perception of what radiation really is results in the absence of natural self-preservation instinct, which is normally present in humans under the real danger. At the same time, hearing information regarding possible exposure to radiation induces radio-phobia in a population with a fear level that is independent from the source of the dose of radiation. For the period of nuclear industry and technology progress (besides the nuclear bombing of Japan) from 1944 until 2001, there were about 420 accidents and catastrophes around the world affecting over 3,000 personnel members of nuclear industry and the population residing in proximity to nuclear facilities.

Chernobyl Accident

The widest scale nuclear disaster took place in Ukraine in 1986 when the unsanctioned power intensification of the reactor occurred followed by its destruction. In that period of time, the power plant staff was limited, and because of secrecy requirements the information about the event was not delivered to all administrative, healthcare, and other services and offices, both within the population residing in NPP satellite of

Pripyat City. It was mostly relevant to the government of Ukraine and local governing bodies. That is why only local arrangements for fire extinguishing were applied, but radiation protection measures for the Pripyat City population and its evacuation were implemented only one day later.

As is well known, 31 people died within several months upon the accident due to radiation injuries and thermal burns [1]. However, the unprecedented arrangements on radiation exposure prophylaxis and radiation impact mitigation were applied to population protection in the vast territories of Russia, Byelorussia, and Ukraine. The Pripyat City population was evacuated within two days, later on i.e., in two weeks people were evacuated from the settlements within 30 km. around the power plant. More than a total of 150,000 people were evacuated [2]. In the short term, extensive work was accomplished with active participation of all republics of the former Soviet Union (FSU) on the "Sarcophagus" construction, which provided conclusion on immediate dose-reducing arrangements being effective in late-1986. Comprehensive check-ups for survivors were provided in almost all healthcare institutions throughout Ukraine [3]. State Registry i.e., the automated system for registration and preliminary health estimate was launched at that time (and is still being utilized today after 19 years) [4].

Radiation burns and acute radiation sickness occurred as the immediate effects of radiation exposure [5]. Such conditions are attributed to early or non-stochastic effects of irradiation. However, probable remote consequences of radiation exposure i.e., so-called stochastic effects (leukemia, radiation cataracts, solid tumors including thyroid cancer) still remain in the field of interest in the global scientific community and scientific institutions [6–8].

According to long-term monitoring of data from the A-bomb survivors in Japan, the extra cases of leukemia had appeared in population within two–five years as a short-term stochastic effect of irradiation. Solid tumors, of various localization, were diagnosed 15–20 years and over [9]. Data, received for the entire survey period, suggests that there were no evidence of genetic abnormalities in offspring generations. Thus, rarely occurring radiation accidents or catastrophes lead to radioactive contamination of territories and provide immediate or remote effects in humans followed by substantial cost, moral and physical expenditures, on those consequences mitigation and liquidation.

Lessons Learned from the 1986 Chernobyl Disaster

First, the reason for the accident was identified and, consequently, the radiation safety measures for nuclear power plants were both applied to population protection for those living in contaminated areas and those close to nuclear facilities. Second, the system of emergent informing was modified for NPP operating regimen, and for cases of nuclear accidents or disasters on various levels including international organizations, and first of all the IAEA. All services and personnel activity coordination system for the accident situation period was launched. Healthcare services of local governing bodies have the required stock for providing emergency aid under the radiation injury both with dosimetry control for the survivors. Finally, the healthcare staff training system based on healthcare institutions and higher medical schools were involved since the accident [11; 12]. A number of scientists and physicians have been trained in specialized centers in Hiroshima, Nagasaki, Moscow, and Kiev. Experts in disaster medicine are trained regularly at IAEA workshops in the field of radiation protection and survivors' management after the radiation injuries.

Information sharing via the Internet

Information exchange through scientific conferences and workshops that utilize Internet resources improve the dissemination of knowledge not only among the radiation experts, but also among the general public [13]. In 2001, our new lecture in Radiation Epidemiology was placed on the Web site of the University of Pittsburgh Supercourse project, outlining general approaches to the field <www.cmkp.edu.pl/~super1/lecture/lec2741/002.htm>. The same year, we launched the Radiation Epidemiology and State Registry for the Chernobyl disaster survivors' Web site in Ukraine <www.is.svitonline.com/epidemua/>, where links to worldwide information sources are provided. These measures advanced not just the spread of extra knowledge, but also an acquired data exchange in the field of nuclear technologies and the possible negative aftermath of radiation impact on humans.

References

[1] Guskova AK, Baranov AE. 1989. Acute radiation sickness diagnostics, clinical pattern and management in the Chernobyl NPP accident survivors. Terapevt. Arch 1:95–103 p.

[2] Ledoshchuk BA. 1991. State distributed registry for exposed to radiation persons after the Chernobyl NPP accident. Vestn. AMN USSR 8:9–15 p.

[3] Ledoshchuk BA. 1991. Targeted check-up system and state distributed registry. Health effects of the Chernobyl NPP accident. Inform. bul. AUSCRM AMS USSR, Kiev 169–178p.

[4] Ledoshchuk BA, Homenko NR. 1994. Actual issues of the national registry of Ukraine functioning. Kiev; Znaniye Publ 39.

[5] Bebeshko VG, Prevarskyj BP, Halyavko IG, Shimelis IV. 1988. Rehabilitation tools for the acute radiation sickness survivors. Proc. Sci. Conf. Health issues of the Chernobyl nuclear power plant accident. Kiev; 161–165p.

[6] Ledoshchuk BA, Gudzenko NA. 2000. Leukemia risk analysis in Chernobyl accident clean-up workers in Ukraine. Proceedings of the 5th International Conference on Probabilistic Safety Assessment and Management, Osaka, Japan, held on November 27-December 1, 2000. (4): 2589-2594 p.

[7] Ledoshchuk BA, Ledoshchuk GM. 2001 December 5-6. Long-term epidemiological study of leukemia among liquidators in Ukraine after the Chernobyl accident. In proceedings of tenth symposium on Chernobyl related health effects, Long-term health effects of Chernobyl accident on liquidators, Tokyo, Japan; December 5-6, 2001. 65-77p.

[8] B. Sobolev WF, Heidenreich I, Kairo P, Jaxob G, Goulko, I. Likhtarev. 1997. Thyroid cancer incidence in The Ukraine after the Chernobyl accident: Comparison with spontaneous incidences. Radiat. Envirom. Biophys 36:195-199p.

[9] Preston DL, Kusumi S, Tokunaga M, Izumi S, Ron E, A.Kuramoto, N. Kamada, H. Dohy, T. Matsui, H. Nonaka, D.E. Thompson, M. Soda, and K. Mabuchi. 1994. Cancer incidence in atomic bomb survivors. Part III: Leucemia, Lymphoma and multiple myeloma. 1950-1987. Radiat.Res. 137 (Suppl), S68-S97 p.

[10] D.E. Thompson, K. Mabuchi, E. Ron, M. Soda, M. Tokunaga, S. Ochikubo, S. Sugimoto, T.Ikeda and M. Terasaki. 1994. Cancer incidence in atomic bomb suirvivors. Part II. Solid tumors, 1958-1987 Radiat.Res 137 (Suppl),S17-S67.

[11] Ledoshchuk B.A., Homenko N.R. Program for workmanship perfection in experts involved in lunching and maintenance of the registries for the Chernobyl disaster survivors provided by the course Chernobyl disaster health aftermath, healthcare providing and social protection of persons subject to registration in the National Registry of Ukraine, MH of Ukraine, SCRM AMS of Ukraine, UCIT & NR of MH of Ukraine, Kiev; 24.

[12] Ledoshchuk BA., Myechov DS., Murashko VA., Ischenko VP, Bychkova NP, Trotsuk NK. 2003. Contemporary approaches in the subject-matter courses planning for the post-graduate training Promeneva diagnostika, promeneva terapiya. 4:71.

[13] Ledoshchuk BA, NK Trotsuk, NG Babkina, Byelyayev Yu. M. 2003. Internet-Supercourse of lectures in epidemiology (Contemporary education methods). Proc. third conf. information technologies in healthcare and applied medicine. May 21–23, 2003, Kiev; 100–103.

Disaster Preparedness and Mitigation: Environmental Health Issues

© 2006 IOS Press. All rights reserved.

The CRESP Amchitka Expedition: A Model for Multi- and Interdisciplinary Research into Radionuclide Contamination of the Marine Environment

Conrad Volz, DrPH, MPH[1a], Charles Powers, PhD [a], Joanna Burger, PhD[c], David Kosson, PhD[d], Michael Gochfeld[e], MD, PhD[c], Barry Friedlander, MD, MPH[e], David Barnes, PhD[f], Lisa Bliss, MLS[g], Stephen Jewett, PhD[f], Mark Johnson, PhD[f], Michael Stabin, PhD, CHP[d], Martyn Unsworth, PhD[h], Vikram Vyas, PhD[e] and Janet Horsch, MLS, MFA[b]

[a]Department of Environmental and Occupational Health, Graduate School of Public Health, University of Pittsburgh, [b]University of Pittsburgh Graduate School of Public Health Center for Public Health Practice, [c]Rutgers, The State University of New Jersey, [d]Vanderbilt University, [e]University of Medicine and Dentistry of New Jersey, [f]University of Alaska Fairbanks, [g]Institute for Responsible Management, [h]University of Alberta

Abstract. This paper argues for an increased focus of multinational research studies to specifically be designed to assess and interpret human and ecological risk aspects of environmental radiation from underground nuclear tests, dumping of radioactive wastes, and release of radionuclides into the marine environment from former and active nuclear weapon production and military facilities. Here, we present a unique, multi- and interdisciplinary model expedition to Amchitka Island, Alaska, to determine any current threat to human health and the environment from radionuclide release into the marine system from underground nuclear tests detonated during the Cold War. Physical methods were used to obtain data to reduce uncertainty regarding transport of radionuclides into the marine environment. Extensive water and sediment sampling and analysis were performed to assure the radiological safety of scientific divers and intertidal gatherers. Radiation survey and personnel monitoring were employed throughout the expedition to quickly identify radiation sources and record individual radiation doses, respectively. Scientists sampled biota at each tropic level, including sedentary and mobile species. Both a screening and in depth radiological analysis of samples for anthropogenic and natural radionuclides was done to determine the safety of human foods, human and ecological risk and establish a baseline of biological data to aid in the development of a long-term stewardship plan. While other expeditions have been conducted in far northern latitudes in connection with above- and underwater

---

[1] Corresponding author: Conrad Daniel Volz, DrPH, MPH, Amchitka Expedition Director, Scientific Director – Center for Healthy Environments and Communities, Assistant Professor of Environmental and Occupational Health, Graduate School of Public Health, University of Pittsburgh, 3109 Forbes Avenue, Suite 210, Pittsburgh, PA, USA 15260. dvolz@eoh.pitt.edu.

nuclear detonations, sunken submarines, and releases from accidents and reprocessing plants, they fail to integrate all procedures necessary to fully evaluate human and environmental risk. Moreover, this expedition is the first to study the effects of underground detonations in northern latitudes providing a model for similar work on Novaya Zemlya, the Russian north latitude marine underground testing site.

Keywords. AISP, CRESP Amchitka Expedition, marine, public health, radionuclide contamination

Introduction Amchitka Island

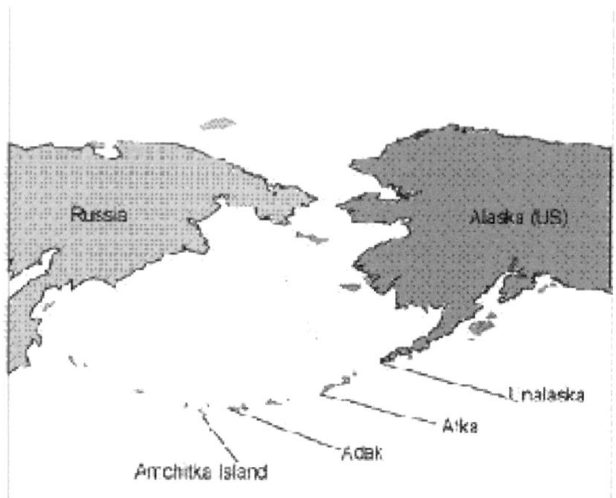

Figure 1. Location of Komandorskiye

Amchitka Island is a member of the Rat Island group of the western Aleutians, located approximately 1200 km east of Petropavlovsk-Kamchatsky and 800 km southeast of the Ostrova (both in the Russian Republic) and 1900 km southwest of Anchorage, Alaska (see figure 1, Location of Amchitka Island). It is located in the most active seismic and volcanic region of the world [1].

Both the Former Soviet Union (FSU) and the United States (USA) have conducted underground nuclear tests on high-latitude islands that are surrounded by complex marine ecosystems, which support large bird nesting areas, high-trophic level marine predators, and commercial fisheries. Between 1964 and 1990, the FSU conducted 43 underground nuclear tests on the Arctic Islands of Novaya Zemlya. The USA detonated three underground thermonuclear devices on Amchitka Island, Alaska—Long Shot (1965, 40-80 kilotons), Milrow (1969, 1 megaton), and Cannikin (1971, the largest underground USA test at approximately 5 megatons). These three test shots account for approximately 16% of all energy releases from the USA underground testing program [2; 3].

A key difference between Novaya Zemlya and Amchitka is that 91 atmospheric tests, with a total yield of approximately 240 megatons (about half the total atmospheric yield), were also conducted on Novaya Zemlya. Additionally, sixteen naval reactors, from seven submarines and the icebreaker *Lenin*, were dumped at five sites in the Kara Sea (an arm of

the Arctic Ocean located east of Novaya Zemlya). The contaminated materials and sources were dumped into the Kara Sea at depths ranging from 12 to 300 meters [4]. Novaya Zemlya designated in 1992 as the Central Test Site of the Russian Republic, constitutes the only open Russian test site. The Kara Sea receives freshwater from the Ob and Yenisei Rivers that carry radionuclides from FSU nuclear production sites, inactive weapons plants, contaminated lagoons and groundwater, and active Russian weapons plants.
The Russian government has proposed using underground sites on Novaya Zemlya as a repository for nuclear wastes throughout their far flung weapons production complex.

While this action has been tabled because of concerns from environmental groups and adjacent countries, the Russian government seems not to have inherent objections to this plan. Numerous nations, including Norway, Japan, Russia, USA, and Canada, conduct fishing operations in these and surrounding areas; and indigenous peoples use these areas for subsistence hunting and fishing. Central governments, state and oblast representatives, and environmental groups express concerns about leakage from these underground nuclear tests for possibly contaminating the surrounding marine environment with radionuclides, thus posing a public health risk from consumption of fish and other marine wildlife.

1. Background of Previous Expeditions to Discover Radionuclide Contamination

There have been numerous multinational and national expeditions and longer term monitoring programs undertaken to determine radionuclide concentrations in water, sediment, and biota in far northern latitudes. Such expeditions have been primarily designed to determine radionuclide levels in response to known releases and to assess human risk or perceived risk from these releases [5] including:

- International Atomic Energy Agency (IAEA), in collaboration with Germany's Bundesforschungsanstalt fur Fischerrei, and the United Kingdom's (UK) Ministry of Agriculture, Fisheries, and Food, to extensively study the Northeast Atlantic dumpsite for nuclear waste.
- Investigation of Plutonium 239 and 240 ratios in the Arctic Ocean from underwater nuclear tests in Chernaya Bay, Novaya Zemlya by J.N. Smith, Bedford Institute of Oceanography in 1993 and 1996.
- Arctic Monitoring and Assessment Program (AMAP), under the lead of Russia and Norway, with assistance from Canada, Denmark, Iceland, the USA, Sweden, and others to the Kara, Barents, and other seas to determine the extent of contamination from the Sellafield Actinide Processing Plant, as well as from sunken FSU nuclear submarines and the Myak Accident in the Southern Urals.
- Norway and Russia have individually and collaboratively studied the environment around sunken nuclear submarines and crashed planes with nuclear warheads in the Barents sea, Pechora, and in the Greenland sea.
- Norwegian marine environmental monitoring includes data from the Norwegian Sea, the North Sea, the Barents Sea, the Skagerrak, and selected fjords in response to the Sellafield Plume movement, as well as FSU sunken submarines, and above

ground and below water tests at FSU Novaya Zemlya test sites. (Friedlander et al. 2005).

The French government has supported research in the South Pacific by the IAEA to assess the potential impact of 147 underground nuclear tests undertaken by France in the Mururoa and Fangataufa Atolls of French Polynesia (IAEA 1998) on the marine environment. However, none of these programs specifically assesses actual radionuclide transport mechanisms to constrain models, bathymetry and marine freshwater anomalies, the on-island depth of the freshwater-saltwater interface, actual levels of radioisotopes in sediment, and different trophic levels of biota, which could bio-accumulate, concentrate, and magnify anthropogenic radioisotopes from below ground nuclear tests. This is a matter of great concern because Novaya Zemlya has been suggested as a deep underground repository for nuclear waste from the vast FSU nuclear waste complex even though it is located in an area that has considerable seismic activity.

Figure 2. Dr. Volz in Bald Eagles Nest

Amchitka Expedition Project Personnel and Objectives

The CRESP Independent Expedition developed and implemented a unique model which incorporated a multi- and interdisciplinary fully-integrated scientific plan to investigate the biological and physical marine environment of Amchitka Island. Expedition planners and advisors, project key personnel, and analytical specialists hailed from diverse scientific and interdisciplinary fields including: oceanography, magnetotellurics, geology and hydrogeology, occupational and environmental medicine, public health, biology, environmental engineering, environmental ethics, radiochemistry, health physics and exposure, and risk assessment sciences.

In the summer of 2004, CRESP embarked on an expedition to Amchitka Island to determine the current threat to human health and the environment of radionuclide release from underground nuclear tests in the surrounding sea waters and to establish a baseline of biological and physical data to reduce uncertainty and aid in the development of a long-term stewardship plan.

2. CRESP the Amchitka Independent Science Plan (AISP) and Community-Based Participatory Research

In February 2002, CRESP teamed up with the University of Alaska- Fairbanks to begin developing an Independent Science Plan, and met with stakeholders to gather concerns and views regarding potential radionuclide contamination emanating from Amchitka Island. CRESP, a multi-university consortium directed by Principal Investigator Charles Powers, PhD, has worked with the United States Department of Energy (DOE) on projects related to radiological and other waste contamination resulting from Cold War legacy nuclear production sites. For over a decade, CRESP has enjoyed the confidence of the DOE and stakeholders regarding the development of conceptual site models linking contaminant movement to receptors, human and ecological risk assessments, hazardous waste remediation, end state land use planning, and communication of risk issues.

During the next 15 months, CRESP, with participation of numerous stakeholders, created an independent, integrated multi- and interdisciplinary scientific plan to investigate if there was breakthrough of radionuclides into the Amchitka marine environment using biological and physical techniques both on land and sea, and to better constrain models of transport and migration [6]. In the summer of 2003, four signatories approved the plan: DOE—National Nuclear Security Agency (NNSA), the Alaska Department of Environmental Conservation (ADEC), United States Fish and Wildlife Service (USFWS), and Aleutian Pribilof Island Association(A/PIA). Planning began for the field expedition in 2004.

The CRESP AISP incorporated a Community-Based Participatory Research (CBPR) approach to determine which foods, actually eaten by subsistence communities, were to be sampled on expedition. This technique has been subsequently shown to provide appropriate human exposure data for risk assessments and the formation of public policy [7]. In the summer of 2003, (JB) and (MG) held meetings with stakeholders, which included Aleut communities on Adak, Atka, Nikolski, and Unalaska Islands, to present the sampling plan and to seek additional information on species important to the nutrition of islanders. Island communities also asked that Aleut hunters/gathers/fishers participate in the expedition—a request that was wholeheartedly endorsed by CRESP, thereby adding to the creditability of the final risk assessments [8].

CRESP Independent Amchitka Expedition

The CRESP expedition was implemented in three phases. Phase I (June 6 to June 23), Phase II (June 28 to July 20), and Phase III (July 18 to August 4).

Figure 3. Lowering Multibeam Sonar

Phase I (June 6 to June 23)

Phase I involved both land and sea physical and radiological safety measurements. Oceanographic activities first included performing a bathymetric survey using an SM2000 multibeam echosounder to determine whether there are features in the bathymetry around Amchitka, such as faults and cavities, that could influence freshwater discharge from the bottom and to ensure that the side scan sonar could be towed safely. The multibeam survey was performed from about 20 to 100 m depth to the northeast of the Cannikin and Long Shot sites. The multibeam sonar had a range scale of 400 meters, interrogated the bottom at 1.92 pings per second, with receiver gain at 35%. This process resulted in bottom mapping to an extremely high 2 m resolution. A Datasonics SIS 1000 side scan sonar mapped the bottom texture over the same area as the bathymetric survey. The instrument has a range scale of 200 meters which produces a 400 meter swath width. Transect lines similar to those of the bathymetric survey were followed while towing the sonar. The resulting images were mapped into a single mosaic. All oceanographic operations were under the direction of Mark Johnson, PhD [9].

Based on at-sea results of the side scan sonar survey, which indicated the presence of sediment, discrete sediment grab-samples were taken. The time, depth, and position of all sediment grabs were recorded. As per the radiation portion of the expedition Health and Safety Plan (HASP), all sediment samples were screened for activity on deck; no activity above background was encountered [10]. Again, to assure the safety of Phase I expedition personnel, and to indicate radionuclides in sediment that might pose a hazard to scientific divers and intertidal collectors in Phase II, a representative sample of all layers present in each grab sample were collected in Nalgene bottles for subsequent gamma spectroscopy analysis at the Vanderbilt Department of Civil and Environmental Engineering. The research team immediately sampled and screened fluid and sediment spillage on the outside of each bottle. All readings were below the minimum detectable activity for the method [11].

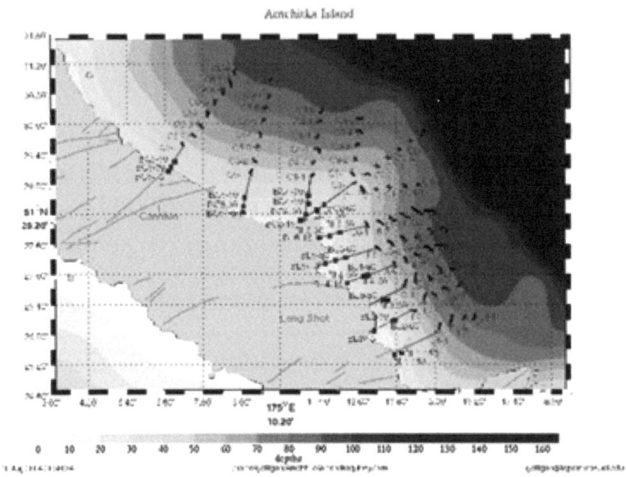

Figure 4. Location of CTD and Dive Sampling Stations, CTD stations in red, dive stations in blue. Fault lines are depicted on Amchitka Island

Conductivity (salinity), temperature, and depth measurements (CTD) were taken to determine evidence of freshwater discharge through the ocean floor, which would indicate possible radionuclide release. A closely spaced CTD survey was completed with emphasis on measuring salinity within the bottom 1 to 2 meters off the ocean floor.

Eighty oceanographic stations along 12 transects were occupied using a Sea-Bird 19 CTD to measure CTD. The CTD was calibrated to salinity accuracy and the altitude of each CTD cast was monitored with an acoustic altimeter.

Bottom seawater samples were taken using a Niskin Sampler at 20 stations on Long Shot and Cannikin transects ranging in depth from 130 to 270 feet. All samples were immediately screened using ambient monitoring and wipe sampling techniques as described above. The findings that all ambient samples were below background radiation levels established in both Adak Island, Alaska and Pittsburgh, Pennsylvania., USA boosted researchers' confidence that gathering biota samples in the intertidal area and underwater by scientific divers could be done without being exposed to radionuclide levels that surpassed expedition radiation guidelines.

The hydrogeology of Amchitka Island, like other small marine islands, is characterized by a layer of freshwater overlying a stagnant saltwater layer. The freshwater is derived from rainfall and snow and offshore discharges. Dependant on freshwater levels and rock porosity, a dynamic balance determines the location of the fresh to saltwater interface (Unsworth et al. 2004). The change from fresh to saltwater typically takes place over a broad transition zone. A key parameter in modeling the timing and quantities of radionuclides that could be released from the test sites on Amchitka Island is the depth of the fresh to salt water transition zone.

Scientists headed by Martyn Unsworth, PhD, University of Alberta, Canada, used Magnetotelluric (MT) measurement, a remote electromagnetic exploration technique, to map the salinity and porosity of the critical sections of the islands subsurface. A major goal was to determine the depth of the fresh-salt water interface at each test shot to develop accurate models with reduced uncertainty regarding the movement of radionuclides from the shot cavity. MT data were collected on profiles that followed transects through each of the test sites. This work was extremely rigorous as personnel had to carry electrodes, car batteries, induction rods, electrical wire, digging implements, and processing stations for long distances over rugged and water-logged tundra. Highly accurate Geographic Positioning System readings were also taken at surveyed standards on Amchitka Island to track island movement relative to the mainland of Alaska. Figure 4 presents the location of MT stations on transect through each test shot

Scientists conducted extensive radiation monitoring of all land activities including dug trenches and holes needed for electrode and rod placement. There was no ambient radiation levels encountered that exceeded background established on Adak Island prior to the expedition departure [12]. Additionally, activity levels decreased as the probe was inserted below ground, as would be expected as a result of ground shielding from radiation of cosmic and cosmogenic origin. David Kosson, PhD, of Vanderbilt University had overall responsibility for coordination and interpretation of the physical aspects of the expedition.

Figure 5. Location of MT Stations

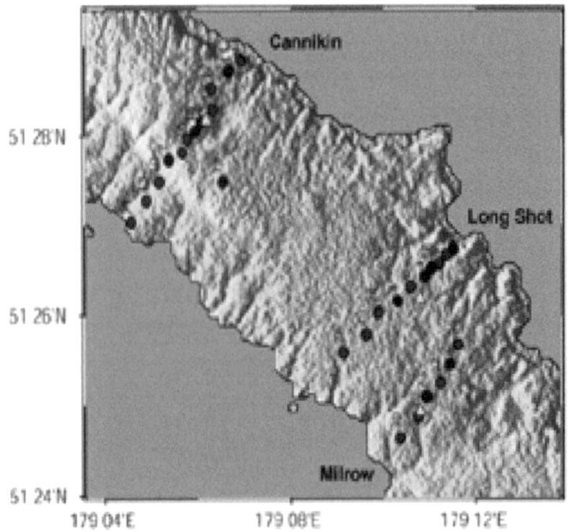

Phase II (June 28 to July 20)

Phase II focused the main expedition purpose—collecting kelp, invertebrates, benthic and higher level trophic-level fish, sea birds, and bald eagle chicks for subsequent radiological analysis. University-based scientific divers sampled biota from subtidal waters. Aleut expedition members used subsistence methods to capture traditional foods including halibut, pacific cod, sculpin, and octopus. Between Phase I and II, biological sampling transects were determined at the geophysical/biological planning conference held on Adak Island. Selected transects occupied for CTD oceanographic work were extended shoreward for dive station sampling, and transects that crossed predicted faults were assigned to these biota sampling stations. In addition to intertidal collections, divers performed subtidal dive collections at 15, 30, 60, and 90 foot isobaths along these transects. Sampling transects were determined for predicted release points from the Cannikin, Long Shot, and Milrow blast sites.

Prior to the expedition, reference sites with marine ecological zones similar to Amchitka Island were determined for comparison purposes, since ocean currents generally flow from west to east and expedition time was limited. Kiska and Buldir islands were considered appropriate controls (although there were confounding variables even with these sites given reports of release of radionuclides into the marine environment from FSU submarine bases and facilities on the Kamchatka Peninsula). Kiska Island was the first reference choice, and even though the expedition encountered severe weather conditions while steaming to this site, the western coast provided ample specimen gathering while waiting for sea conditions to calm before collecting in Kiska Harbor on the east coast, the predetermined optimal gathering location.

The finding of no fissile material or products in Phase I sampled sediment was reported to Dr.Volz on route to Amchitka for this phase of the expedition. Water and sediment samples taken by scientific divers, as well as all specimens taken on either Amchitka or Kiska Islands showed no activity over background. Additionally all wipe sampling of tables after dissection of specimens taken in marine areas near the test sites showed no activity over the minimum detection limit of the method [13].

Phase III (July 18 to August 4)

Phase III involved stationing a technician on a commercial trawler, the *Gladiator,* operated by the U.S. National Oceanographic and Atmospheric Administration (NOAA) to collect fish and crustaceans for radiological analysis. This biennial survey trawls at exact locations to determine the species and sizes of fish present in order to establish the health of the fishery. The survey was conducted on the northern Pacific Ocean side, off reference sites, as well as off the southern coast of Amchitka Island. Species taken for analysis during this survey included Pacific Cod, Atka Mackerel, Halibut, Rock Sole, and Rockfish.

3. Biological Sampling Overview

The actual biological sampling performed in Phase II and III, followed the initial design, and two key objectives were achieved: 1) having sufficient samples of some species in each study area to allow for an adequate statistical comparison, and 2) obtaining a few samples from a range of key species surrounding Amchitka that serve as a base for food chains, ultimately leading to humans. Species that fall into category1 include green sea urchin, kelp, rock greenling, pacific cod, gulls, eiders and puffins. Category 2 species include Atka mackerel, halibut, octopus, eagle and a single Stellar Sea Lion taken by Aleut hunters under their subsistence hunting rights, granted them by treaty. Other very important features of the collection phases were that the team was able to collect Aleut subsistence foods in the intertidal beaches near the test shots and at the reference site including mussels, chitons and limpets; and the diversity of ages of biota collected were well represented. Kelp, sea urchins and jingles are not long-lived, while some of the fish collected can reach 50 to 100 years of age in this ecosystem [14]. All biological sampling was performed under the direction of Joanna Burger, PhD.

Chain of custody forms followed specimens through biological preparation and analysis steps. A total of 45 coolers, each appropriately sealed, numbered and with attached chain of custody control forms were sent from Adak Island to Newark Airport. These coolers contained over 3,000 pounds of specimens.

4. Specimen Testing Plan

The radioisotopes of interest for analysis of biota were ^{137}Cs, ^{152}Eu, ^{60}Co, $^{238, 239, 240, 241}Pu$, $^{234, 235, 236, 238}U$, ^{241}Am, ^{90}Sr, ^{3}H, ^{99}Tc and ^{129}I. ^{137}Cs and ^{90}Sr are the isotopes most likely to accumulate in soft tissue including muscle and cause human health risk through consumption. The other isotopes expected to be released as a result of a nuclear detonation accumulate preferentially in either skeletal material (bones or exoskeleton) or specific organs.

The CRESP analytical team employed a dual screening analysis of a wide range of organisms representing different trophic levels to understand the occurrence and give an indication of the origin of these radionuclides. The Idaho National Engineering Laboratory (INEL) screened soft tissue, bone and kelp for a broad range of alpha, beta and gamma emitters, followed by a focus on alpha emitters in bone and kelp. Specimen testing for ^{137}Cs, other relevant gamma emitters and interlaboratory comparisons was done at Vanderbilt University. Analysis of samples at Vanderbilt University was done under the supervision of Michael Stabin, PhD, CHP. At Vanderbilt University a broad range of Aleut foods and food chain organisms were analyzed and there was an increased focus on analysis of a wider spectrum of each species and locations, more samples of selected species to achieve statistical power and analysis of larger sample sizes (1000 grams) for longer count times to lower the detection limits for ^{137}Cs [15].

5. Summary

All nations and supra-governmental organizations have a stake in the suitability, reliability, and sustainability of environmental monitoring and engineering and institutional controls at former Cold War test sites, nuclear production sites and active sites in both Russia and the USA. There exists a catastrophic potential of nuclear contamination to the Kara Sea and Arctic Ocean from legacy wastes at former FSU nuclear production facilities located in the drainage systems of the Ob and Yenisei rivers. This material if released due to seismic and/or weather related phenomena or terrorist attack, could potentially affect fisheries on and off the continental shelves of Europe, Asia, and North America.

Release of this material in aqueous and/or aerosol form could also potentially cause a public health emergency within Russia necessitating the evacuation of hundreds of thousands of people.

While there have been numerous multinational research expeditions to assess and characterize the risks posed to marine resources from underground nuclear tests, dumping of radioactive wastes, and release of radionuclides into the marine environment from former and active nuclear weapon production and military facilities, these studies often fail to integrate all procedures necessary to evaluate human and environmental risk. The Amchitka expedition model was specifically designed to assess and interpret the human and ecological risk aspects of environmental radiation. Additionally, this was the first expedition to study the effects of underground detonations in northern latitudes and could provide a model for similar work on Novaya Zemlya, the FSU underground testing site.

The Supercourse and Contamination from Legacy Cold War Military Production

Researchers, lecturers, and students at schools of public health, medicine, nursing, environmental health, engineering, and public and international ffairs, governmental and international agencies, and professional and environmental groups around the world must have a forum for discussion of issues related to contamination as a result of Cold War testing and military and industrial processes. The Supercourse Web site receives approximately 75,000,000 hits per year making it an excellent vehicle for distribution of information, data, lectures and alerts relative to cross-boundary issues inherent in understanding and controlling Cold War Legacy contamination. Lectures and reports regarding the status of remedial actions and residual contamination at these critical sites could be circulated worldwide via the Supercourse.

6. Acknowledgements

This expedition and the safe return of all participants would not have been possible without the skill and dedication of Jerry Downing of B&N Fisheries and the captain and crew of the *F/V Ocean Explorer*. We thank all field scientists and technicians for long hours of backbreaking work under extremely harsh weather conditions that were essential to the success of the expedition. The hunting and gathering of native foods could not be performed in subsistence fashion without the wisdom, skill, and instincts of our Aleut team members. Our thanks to James Weston of the University of Mississippi for gathering fish specimens as part of the NOAA trawl survey; to Anne Morkill and Vernon Byrd of the U.S. Fish and Wildlife Service who gave important input into our knowledge of both land and marine conditions at Amchitka Island, as well as permitting help. (CV) acknowledges Bernard Goldstein, MD, Dean of the University of Pittsburgh's, Graduate School of Public Health for his guidance over the past two years, and also thanks Faina Linkov, PhD, and Ron LaPorte, PhD, of the Supercourse for their invitation to speak at the NATO conference and for their logistical assistance.

This research was funded by the Consortium for Risk Evaluation with Stakeholder Participation (CRESP) through the Department of Energy (AI # DE-FC01-95EW55084, DE-FG 26-00NT 40938). (CV) is also funded through an environmental health grant from the Heinz Endowment through the Graduate School of Public Health, Center for Healthy Environments and Communities.

References

[1] Wood CA, Kienle J. 1990. Volcanoes of North America: United States and Canada. New York: Cambridge University Press.

[2] Robbins A, Makhijani A, Yih K. 1991. Radioactive heaven and earth – the health and environmental effects of nuclear weapons testing in, on, and above the earth. New York: Apex Press.

[3] DOE (Department of Energy).2000. United States Nuclear Tests July 1945 through September 1992. DOE/NV-209, Nevada Operations Office, Las Vegas, Nevada.

[4] Mount ME, Layton DW, Lynn NH, Hamilton TF. 1998. Use of (59)Ni and (236) U to monitor the release of radionuclides from objects containing spent nuclear fuel dumped in the Kara Sea. Internat. Symp. Mar. Poll., Monoco. US DOE W-7405-ENG-48.

[5] Friedlander B, Burger J, Gochfeld M. 2005. A Review of the radionuclides in the marine environment, CRESP.

[6] Burger J, Gochfeld M, Kosson D, Powers C, Friedlander B, Eichelberger J, Barnes D, Duffy L, Jewett S, Volz C. 2005. Science, policy and stakeholders: developing a consensus science plan for Amchitka Island, Aleutians, Alaska. Environmental Management 35(5): 557-568.

[7] Suk WA, Avakian MD, Carpenter D, Groopman JD, Scammell M, Wild CP. 2004 February. Human exposure monitoring and evaluation in the Arctic: the importance of understanding exposures to the development of public health policy. Environmental Health Perspectives 112(2):113-20.

[8] Powers C, Burger J, Kosson D, Gochfeld M, Barnes D (ED), Powers C, Burger J, Kosson D, Gochfeld M, Barnes D, Bliss L, Friedlander B, Jewett S (PhD), Johnson M, Stabin M, Unsworth M, Volz C, Vyas V, Weston J. 2005 August 1. Amchitka independent science assessment: biological and geophysical aspects of potential radionuclide exposure in the Amchitka Marine Environment, Anchorage Alaska.

[9] Johnson M and Stewart. 2005. Results from the Amchitka Island oceanographic study, CRESP.

[10] Volz (a) C. 2005. CRESP: Amchitka expedition, radiation survey monitoring report.

[11] Volz (b) C. 2005. CRESP: Amchitka expedition anti-cross contamination procedures.

[12] CRESP 2004. Amchitka biological specimen assessment. An independent analysis to be conducted by CRESP (as lead project manager), Rutgers University, Vanderbilt University, Idaho National Laboratory and the University of Pittsburgh.

[13] Joanna Burger, Michael Gochfeld, Sean Burke, Christian W. Jeitner, Steven Jewett, Daniel Snigaroff, Ronald Snigaroff, Tim Stamm, Robert Patrick, Conrad D. Volz and James Weston. 2005. Do scientists and fishermen collect the same size fish: implications for risk assessment. CRESP.

[14] CRESP 2002. Proceedings of the Amchitka Island long-term stewardship workshop. 2002 February 12-14. Consortium for risk evaluation with stakeholder participation. Fairbanks, AK: University of Alaska, Fairbanks.

[15] CRESP. 2003. Amchitka independent assessment science plan. Consortium for risk evaluation with stakeholder participation, June 2003. Available from: http://www.cresp.org/amchitka.html

[16] CRESP Web site. Available fro: http://www.cresp.org/

[17] Unsworth, M, Soyer W, Tuncer V. 2005. Magnetotelluric measurements for determining the subsurface salinity and porosity structure of Amchitka Island, Alaska. CRESP.

© 2006 IOS Press. All rights reserved.

Respiratory Symptoms Reported by Adults Living in an Air Polluted Area in Jordan

Kawkab SHISHANI, PhD, MSN, BSN
Assistant Professor
The Hashemite University

Abstract. The Hashemite Kingdom of Jordan is a small country in the Middle East (89,342 sq. km) with a population of only about 4.8 million (1998). Exposure to air pollution aggravates respiratory diseases. Kuo, Lai, Lee, Tai, and Lee (2002) found a positive relationship between prevalence of asthma diagnosed by lung function test and air pollution. In the current study lower respiratory tract symptoms were prevalent more than upper respiratory tract symptoms. No study was found to distinguish the prevalence rates of respiratory symptoms in this manner.

Keywords. Jordan, pollution, public health, respiratory

Introduction

The Hashemite Kingdom of Jordan is a small country in the Middle East (89,342 sq. km) with a population of only about 4.8 million (1998).However, with an annual growth rate of 3.3 %, Jordan's population is increasing rapidly. It is estimated that if the rate of growth does not decline, the population of Jordan will double again within the next 19 years. An overwhelming majority (87.7%) of the population lives in urban areas.

Air pollution in Jordan is rising. According to the Royal Scientific Committee (2001), Al-Hashemeyia area is one of the most air-polluted areas in Jordan. Sources for air pollution in Al-Hashemeyia area are: 1) Stationary sources including the oil refinery, King Hussein Thermal Power Plant (an oil power plant), and Kherbat Assamra landfills. Some of the hazardous chemicals produced by these sources are sulfur dioxide (SO_2), carbon monoxide (CO), carbon dioxide (CO_2), nitrogen oxides (NOx), and particulate matter (PM_{10}); 2) mobile sources related to combustion-engine vehicles such as diesel and gas-powered vehicles are significant air pollution sources in this area since the area is crossed by two highways with heavy traffic [5]. In fact, levels of CO, H_2S, and NO in Al-Hashemeyia, measured monthly over a one year period and taking into consideration the weather parameters, exceeded the permitted levels [17.]

1. Literature Review

Air pollution originates from human-made sources, including mobile sources (e.g., cars, trucks, buses) and stationary sources (e.g., factories, refineries, power plants). Breathing contaminated air is one of the ways that can pose health risks. In fact, countries such as Belgium, the United States of America, and United Kingdom witnessed a significant rise in the numbers of morbidity and mortality due to high levels of air pollution [2; 4].

People exposed to toxic air pollutants at sufficient concentrations and durations may have an increased chance of experiencing serious health problems. In fact, animal experiments demonstrated that rats exposed to air pollutants developed a marked airspace inflammatory response because of retaining pollutants in the interstitial tissue of the lung [18]. In humans, air pollutants change chemical reactions within individual cells; these changes impair cell function and eventually organ damage [5].

Prevalence of respiratory symptoms is common among people exposed to air pollutants. Two border cities in the U.S. and in Mexico were compared to each other in terms of prevalence of respiratory symptoms. Selected air pollutants such as PM in the Mexican city was higher (104 $\mu g/m^3$) compared to the U.S. city (57 $\mu g/m^3$) ($p < 0.001$). There was significant difference in the overall respiratory symptoms reported in the Mexican city (38%) and the American city (24%). Prevalence of nasal symptoms rate in the U.S. city was 18% compared with 30% in the Mexican city ($p < 0.001$), cough prevalence rates were 13% and 18%, respectively ($p < 0.001$), and daily sputum production was 2% and 9%, respectively ($p < 0.001$) [20].

Exposure to air pollution aggravates respiratory diseases. Kuo, Lai, Lee, Tai, and Lee (2002) found a positive relationship between prevalence of asthma diagnosed by lung function test and air pollution ($r = .63$). In a case-control study, based on lung function test, there was an increase risk of asthma (OR = 1.72, CI = 0.71-4.17) among workers exposed to sulfur compounds at workplace [7].

Exposure to air pollution has been associated also with symptoms of cough, phlegm, and wheezing without having a cold [23], respiratory infections [15] increase in chronic obstructive pulmonary disease (COPD) emergencies [21]. Moreover, Montnemery and colleagues (2003) demonstrated that 36% of people living in polluted area reported upper respiratory symptoms compared to 31% living in non-polluted area ($p <.05$). Living in a polluted area was a risk factor for clinical diagnosis of chronic bronchitis or emphysema (OR = 2.6, CI = 1.9-3.6).

Mortality is also an outcome of air pollution. In fact, it was associated with increase in daily total number of deaths [8; 9]. In several European cities, sulphur dioxide (SO_2) was associated with a 3% increase (CI = 2%-4%) in daily mortality rate and for PM was 2% increase (CI = 1%-3%) [9]. A meta-analysis that examined the effect size estimated in daily mortality and airborne particles concentration. The analysis demonstrated that airborne particles (PM) concentration was a significant risk factor for elevated mortality (RR = 1.06, CI = 1.05-1.07) [19.]

Exposure to air pollutants affects lung function. Pulmonary function tests were completed on 3,293 subjects. PM, O_3, and NO_2 were significantly associated with decrease in pulmonary function tests with regression coefficients (B) of -32.0 ($p < 0.05$), -94.2 ($p < 0.01$), -42.6 ($p < 0.05$) respectively [16.] Furthermore, Al-Neaimi, Gomes, and Lloyd

(2001) demonstrated that ventilatory lung functions were lower for people exposed to cement dust.

Air pollution poses a threat to pregnancy outcomes. Maisonet, Bush, Correa, and Jaakkola (2001) evaluated the relationship between term low-birth weight and CO, PM, and SO. Exposure to CO was associated with increase the risk for low birth weight (OR = 1.31, CI = 1.06-1.62).

Purpose

For the last three decades, environmental problems in Jordan, are rising [14]. However, no data is available on the health status of residents of Al-Hashemeyia area. Therefore, this is the first epidemiological study to describe prevalence of self-reported respiratory symptoms by adults18-years and older living in Al-Hashemeyia area, and examine the relationships between characteristics of the sample and self-reported respiratory symptoms.

2. Methodology

Research Design

This study utilized a descriptive design. Participants were recruited from Al-Hashemyia area. The Hashemite University where the primary investigator of this research is employed is located 8 kilometers from Al-Hashemyia area. This suburban town is part of the Zarqa district in Jordan. The total population of the area is 13,000 [3.]

Sample

The target population for this study is composed of all adults living in the Al-Hashemeyia for at least one year. Eligibility criteria include that: (1) participants must be 18-years and older and (2) participants must be a resident of Al-Hashemeyia for at least one year.

The study sample was computed using PASS 2002 power analysis and sample size software. In a cohort study with no background incidence of a particular adverse reaction, the probability that one or more adverse reaction will not occur in a sample of 400 individuals is 20%; the power of the study is 80%.

The Instruments

Demographic

Demographic data were collected as part of the Respiratory Risk Profile on age, gender, employment, level of education, income, address (neighborhood), smoking behavior, length of residence in Al-Hashemeyia, history of asthma and related medications, and daily length of time spent outdoors in Al-Hashemeyia area. Additional data were collected on personal and family history of cancers and any other health problems, and people's concerns regarding the quality of air in Al-Hashemeyia. Subjects were asked about presence of eye symptoms (watery, itchy, red eyes) and skin symptoms (eruptions, eczema).

Respiratory Risk Profile (RRP)

This instrument was developed by Shishani to evaluate the prevalence and severity of respiratory symptoms. The development of this instrument came as a response to the unavailability of such instrument in the Arabic language. The Respiratory Risk Profile was developed through a synthesis of international studies that examined outdoor air pollution effects on health. The instrument is self-report and consists of 16 items measuring symptoms directly related to respiratory symptoms during the past three months. The instrument is grouped into upper and lower respiratory symptoms. All items are worded positively to allow one direction for all items. That is, every item was scored so that higher scores meant higher potential for respiratory symptoms. Responses are based on a 5-point Likert scale ranging from 1 = never, rarely = 2, sometimes = 3, most of the time = 4, and always = 5.

The instrument in its totality was pilot-tested among 100 respondents from the target population, and the preliminary testing has shown an acceptable level of internal consistency (Cronbach's Alpha = 0.93). An expert panel of five professionals evaluated content validity: three pneumologists, a respiratory therapist, and a researcher with an expertise in survey instrument development.

Item analysis was performed to assess each item's contribution to the total scale by evaluating item-total correlations. The item-total correlations ranged from .46-.78. This means that there were strong relationships between the individual items and total scores. The item-total correlations demonstrated that all of the items are contributing to the measurement of the construct (respiratory symptoms).

Procedures

The human subjects' approvals were obtained from the Hashemite's University Scientific Board. Residents of Al-Hashemeyia who met the eligibility criteria of the study were asked to participate in this study through project advertisements in the Ministry of Health (MOH) primary health centers and the municipality offices, and through word-of-mouth using a door-to-door approach. Four research assistants distributed a total of 400 copies of the instrument. A cover letter indicating the purpose of the study, confidentiality statement, and the primary researcher contact information accompanied the instrument.

Research assistants were nurses with master's degree in clinical nursing. They all received a one-day training by the primary investigator to ensure standardization of data collection procedure. Training included handling possible requests from subjects regarding explanations of certain items or words. Research assistants explained the study and its significance and all foreseeable benefits/risks. Because of cultural sensitivity to gender issues, male and female research assistants formed a data collection team, where the male research assistant collected data from male subjects while the female research assistant approached and intervened with the female subjects. Most of the subjects preferred to complete the requested information at the same time they received the package. Less than 5% of the subjects completed the instrument within few days; those subjects called the primary researcher at workplace asking to send a research assistant to pick up the instrument.

3. Results

Statistical analysis procedures were performed using SPSS for windows version 11.5 (Chicago, Illinois) at a significance level of .05. Descriptive statistics were used to generate means, standard deviations, and frequencies for self-reported respiratory symptoms. Correlations were used to examine the relationships between demographics and self-reported respiratory symptoms.

Sample characteristics

Descriptive statistics were used to summarize characteristics of the sample (Table 1). Respondents were men (76%) and women (24%) with a mean age of 34 (SD = 13). Approximately half of the sample completed up to high school. Participation was slightly higher among nonsmokers (59%) than smokers (41%). The majority of the subjects lived in Al-Hashemyia all their lives, and spent outdoors a daily average of 6 hours (SD = 5).

Table 1. *[cont. on next page]* Sample characteristics (n= 400)

Characteristic	**N (%)**
Gender	
Male	303 (76)
Female	95 (24)
Marital status	
Single	112 (28)
Married	240 (60)
Other	48 (12)
Education	
Illiterate	26 (7)
High school	180 (45)
College	157 (39)
Graduate	16 (4)
Income	
Low (less than $ 300)	92 (23)
Medium ($300-600)	268 (67)
High (more than ($600)	19 (5)
Employment	
Unemployed	50 (11)
House wife	33 (9)
Skilled worker	22 (5)
Retired	43 (10)
Other	242 (65)
Smoking status	
Yes	160 (41)
No	235 (59)
Number of cigarettes smoked daily	
Less than 10	9 (6)
10-20	88 (58)
more than 20	55 (36)

Age	
Less than 25	124 (31)
25-34	111 (28)
35-44	74 (19)
45-54	48 (12)
More than 55	43 (10)

The total number of self-reported cancer cases was 137. About 15% of the cancer cases (n= 24) were family members and the rest were relatives from first generation. The overall distribution of cancers were leukemia (26%, n= 35), lung cancer (12%, n= 16), brain cancer (8%, n= 11), breast cancer (7%, n= 10), prostate cancer (4%, n= 5), uterus cancer (3%, n= 4), and bone cancer (3%, n= 4).

In addition, subjects reported symptoms related to eyes and skin. These symptoms were watery eyes (39%, n= 149), itchy eyes (38%, n= 146), red eyes (31%, n= 117), skin eruptions (13%, n= 13), and eczema (11%, n= 41).

Residents of AL-Hashemeyia expressed several concerns to the research assistants. They reported that symptoms were relieved when they were away from AL-Hashemeyia. Further, they were concerned regarding the quality of air in AL-Hashemeya, and they were confident that air plays major role in health. Furthemore, AL-Hashemeya residents were dissatisfied with the current quality and quantity of health services available in their area.

Prevalence of self-reported respiratory symptoms

Table 2 summarizes the means, standard deviations, frequencies, and %ages of self-reported respiratory symptoms during the previous three months. On a 5-point Likert scale, the total mean of self-reported respiratory symptoms was 3.20, which indicates that respiratory symptoms occurred sometimes for the past three months. In descending order, the highest means were for deep bronchial cough (Mean = 3.7), and dry cough (Mean = 3.5).

Table 2 *[cont. on next page].* Means, standard deviations, and frequencies of the self-reported respiratory symptoms (N=400)

			N (%)	N (%)	N (%)	N (%)	N (%)
Respiratory symptoms	μ	S.D.	1	2	3	4	5
Difficulty in breathing when inhale dust or fumes	2.5	1.4	61 (15)	43 (11)	84 (21)	65 (17)	140 (36)
Flu lasting more than one week	2.7	1.4	70 (18)	43 (11)	77 (20)	77 (20)	117(31)
Difficulty in breathing when exercising	2.7	1.4	60 (15)	54 (14)	86 (22)	75 (19)	116 (23)
Difficulty in breathing because of nasal congestion	3.0	1.4	85 (22)	46 (12)	108 (27)	74 (19)	79 (20)
Mucus without having cough	3.4	1.4	88 (22)	57 (14)	90 (25)	76 (20)	71 (19)
Difficulty in breathing when exposed to cold air	3.1	1.4	92 (24)	63 (16)	89 (23)	71 (18)	73 (19)
Itchy nose	3.1	1.3	85 (22)	69 (18)	100	57 (15)	68 (18)

					(27)		
Difficulty in breathing	3.2	1.3	97 (25)	73 (19)	93 (24)	63 (17)	56 (15)
Pain in sinuses	3.4	1.4	113 (30)	74 (20)	83 (21)	51 (14)	55 (15)
Itchy throat	3.2	1.3	90 (24)	57 (15)	120 (32)	54 (15)	52 (14)
Runny nose	3.3	1.3	104 (27)	69 (18)	108 (28)	57 (15)	47 (12)
Chest wheezing	3.4	1.3	112 (29)	79 (21)	90 (23)	51 (13)	52 (14)
Mucosy cough	3.3	1.3	100 (27)	69 (18)	108 (27)	54 (15)	46 (13)
Dry cough	3.5	1.3	109 (28)	89 (23)	90 (24)	52 (14)	40 (11)
Hoarseness	3.4	1.2	90 (24)	81 (22)	120 (31)	44 (12)	41 (11)
Deep bronchial cough	3.7	1.3	153 (40)	73 (19)	76 (20)	43 (14)	34 (9)

The symptoms were categorized, according to severity, into five groups (never =1, rarely =2, sometimes =3, most of the time =4, and always =5). Symptoms were considered severe when reported always or most of the time. The most frequent respiratory symptoms reported, most of the time and always, were difficulty in breathing when inhale dust or fumes (53%), flu lasting more than one week (51%), difficulty in breathing when exercising (42%), difficulty in breathing because of nasal congestion (39%), mucus without having cough (39%), difficulty in breathing when exposed to cold air (37%).

Relationships between demographics and self-reported respiratory symptoms

Using correlations, demographics (gender, use of asthma medication, smoking, age, hours spent daily outdoors) were examined in relation to self-reported respiratory symptoms. Correlations were done based on the total score of reported respiratory symptoms. According to table 3, only one factor was found to be significantly related reported respiratory symptoms. This factor was use of asthma medication (r = 0.27, p = .01).

Table 3. Correlations between demographics and self-reported respiratory symptoms

	Gender	Asthma medication	smoking	age	Hours spent outdoors	Respiratory symptoms
Gender	1					
Asthma medication	-.06	1				
smoking	.33	-.01	1			
age	-.02	.04	.07	1		
Hours spent outdoors	-.20	.02	.08	-.20	1	
Respiratory symptoms	-.01	.27**	.03	-.01	-.05	1.0

** p = .01

4. Discussion

This study examined the prevalence of respiratory symptoms in adults living in one of the most air polluted areas in Jordan, and the characteristics associated with self-reported respiratory symptoms. This is the first research done to assess health status of residents of Al-Hashemyia area. In the current study, lower respiratory tract symptoms were prevalent more than upper-respiratory tract symptoms. No study was found to distinguish the prevalence rates of respiratory symptoms in this manner. A thorough investigation was done to find statistics on prevalence of respiratory health problems in Jordan in order to make comparisons with the current study findings. However, no relevant information was found within the Ministry of Health and the Department of Statistics Databases.

Therefore, the frequencies of self-reported respiratory symptoms in this study were compared with those self-reported in international studies that examined health problems of people living in polluted areas. The prevalence rates of respiratory symptoms in the current study were similar to the rates reported by Al-Neaimi et al. (2001); Marttila et al. (1994); Montnemery et al. (2003); Stephen et al. (2003), and were slightly higher than the prevalence rates reported by Zulkarnain et al. (2003).

Demographics (gender, smoking, etc.) were not significantly related to respiratory symptoms. Similar results were reported in the literature. Stephen and colleagues (2003) and Montnemery et al. (2003) demonstrated that gender was not linked to self-reported respiratory symptoms. Using objective measure Kuo et al. (2002) concluded that poor lung function tests were similar in male and female subjects living in an air polluted area.

Smoking in this study was not associated with respiratory symptoms. Montnemery et al. (2003) examined the relationship between smoking and respiratory symptoms and reported that smoking did not have any impact on prevalence of nasal symptoms. Al-Neaimi and colleagues (2001) demonstrated that the mean differences in the pulmonary function values among the exposed smokers and non-smokers and among the unexposed smokers and non-smokers was not statistically significant at 95% confidence.

Only use of asthma medication was significantly related to respiratory symptoms in the current study. Kuo et al. (2002) and Jaakkola et al. (2003) reported positive relationship between prevalence of asthma diagnosed by lung function test and air pollution. On the other hand, Al-Neaimi and colleagues (2001) did not find differences in reported asthma between exposed and unexposed workers to cement dust. Moreover, Montnemery et al. (2003) demonstrated that respiratory symptoms were not linked to asthma (OR = 1.0, CI = 0.6-1.5).

Limitations of the study

This study has several limitations. The study used convenience sampling technique. Thus, results cannot be generalized beyond the sample. The study sample was adults living in one of the most air polluted areas in Jordan. Using a comparison group from air in a non-polluted area would have been helpful to know whether the prevalence of respiratory symptoms in Al-Hashemyia is high.

In addition, there is a possibility of underestimation in self-report respiratory symptoms. There was consensual agreement among the individuals who participated in the current study about two issues. First, many of the subjects expressed that there is no need to do a study to examine the prevalence of respiratory health problems because they all are suffering from the quality of air in AL-Hashemyia. The second issue is that air pollution has been the case for more than four decades and it is getting worse because of lack of officials' interest in the area, therefore, the subjects expressed that no one would care about what AL-Hashemyia residents report regarding their health problems.

Furthermore, it would have been ideal to measure lung function tests and correlate it to the Respiratory Risk Profile to examine concurrent validity. Finally, this study did not examine the relationships between air pollutants and respiratory symptoms. Doing so would have strengthened the causal inferences regarding the relationship between air pollution and respiratory health problems. However, for practical reasons this was impossible in the current study.

References

[1] Al-Neaimi Y, Gomes J, Lloyd O. 2001. Respiratory illnesses and ventilatory function among workers at a cement factory in a rapidly developing country. Occup. Med 51(6):367-373p.
[2] Bell M, Davis D. 2001. Reassessment of the lethal london fog of 1952. Environmental Health Perspectives 109 (sup 3), 389-394p.
[3] Department of Statistics. 2005. [Accessed January 20, 2005] Available from: http://www.dos.gov.jo
[4] Dockery D, Pope C, Xu X, Spengler J, Ware I. 1993. An association between air pollution and mortality in six US cities. New England Journal of Medicine (329):1753-1759p.
[5] Environmental Protection Public Agency. 1991. National strategy for environmental protection in Jordan. Jordan: Adustor Press.
[6] EPA 2005. Risk assessment for toxic air pollutants: A citizen's guide. [Accessed January 10, 2005] Available from: http:// www.epa.gov/cgi-bin/epaprintonly.cgi
[7] Jaakkola JJ, Piipari R, Jaakkola MS. 2003. Occupation and asthma: a population-based incident case-control study. American Journal of Epidemiology 158(10):981-7p.
[8] Jelinkova J, Branis M. 2001. Mortality during winter smog episodes 1982, 1985, 1987 and 1993 in the Czech Republic. *International* Archives of Occupational & Environmental Health 74(8):565-73p.
[9] Katsouyanni, K, Toulomi G, Spix C, Schwartz J, Balducci F, Medina S, Rossi G, Wojtyniak B, Sunyer J, Bacharova L, Schouten J, Ponka A, Anderson H. 1997. Short term effects of ambient sulfur dioxide and particulate matter on mortality in 12 European cities: results from time series data from the APHEA project. BMJ 314(7095):1658-1663p.
[10] Kuo H, Lai J, Lee M, Tai R, Lee M. 2002. Respiratory effects of air pollutants among asthmatics in central Taiwan. Arch Environ Health 57:61-68p.
[11] Maisonet M, Bush T, Correa A, Jaakkola J. 2001. Relation between ambient air pollution and low birth weight in the northeastern United States. Environmental Health Perspective 109 (3):351-356p.
[12] Marttila O, Jaakkola, J., Vilka, V, Jappinen, P, Haahtela T. 1994. The south Karelia air pollution study: the effects of malodorous sulfur compounds from pulp mills on respiratory. Environ. Res 66:152-159p.
[13] Montnemery P, Popovic M, Andersson M, Greiff L, Nyberg P, Lofdhal C, Svensson C, Persson C. 2003. Influence of heavy traffic, city dwelling and socioeconomic status on nasal symptoms assessed in a postal survey. Respir Med 97:970-977p.
[14] Mrayyan B. 2002. An overview of the environmental pollution problems in Jordan and identifications of the main polluted sites. [Accessed November 1, 2003] Available from http://www.ics.trieste.it/documents/chemistry/remediation/activities/egm-Apr2002/Jordan-full.pdf
[15] Pariti -Pellinen K , Martilla O, Vilkka V, Jakkola J, Jappinen P, Haahtela T. 1996. The South Karelia air pollution study: effects of low-level exposure to malodorous sulfur compounds on symptoms. Arch Environ Health 51 (4), 315-320p.

[16] Peters J, Avol E, Edward W, Gauderman J, Linn W, Navidi W, London S, Margolis H , Rappaport E , Vora, H, Gong H, Thoma D. 1999. A study of twelve southern California communities with differing levels and types of air pollution. Am J Respir Crit Care Med 159:768-775p.
[17] Royal Scientific Society 2001. Study of the quality of air in Al-Hashemyia-Zarqa March 2000-March 2001. Royal Scientific Society: Center for Environmental Research.
[18] Seaton, A. MacNee, W. Donaldson, K, Godden, D. 1995. Particulate air pollution and acute health effects. The Lancet 345(8943):176-178p.
[19] Schwartz J. 1994. Air pollution and daily mortality: A review and meta analysis. Environmental Research 64:36-52p.
[20] Stephen G, McRill C, Mack M, O'Rouke M, Flood T, Lebowitz M. 2003. Assessment of respiratory symptoms and asthma prevalence in a US-Mexico border region. Arch Environ Health 8 (3):156-162p.
[21] Tenias M, Ballester F, Perez-Hoyos, Rivera M. 2002. Air pollution and hospital emergency room admissions for chronic obstructive pulmonary disease in Valencia, Spain. Arch Environ Health 57:41-47p.
[22] Touloumi G, Samoli E, Katsouyanni K. 2002. Daily mortality and winter-type air pollution in Athens, Greece a time series analysis within the APHEA project. Journal of Epidemiology and Community Health, 50 (suppl 1), s47-51.
[23] Zulkarnain D, Sudarmadi S, Suzuki S, Kawada T, Tomoyuki T. (2003). Effects of air pollution on respiratory health in Indonesia and its economic cost *Arch Environ Health, 51* (4), 135-143.

Scientific Networking and the Global Health Network Supercourse
F. Linkov and R. LaPorte (Eds.)
IOS Press, 2006
© 2006 IOS Press. All rights reserved.

The 1999 Marmara Earthquake and Afterwards[1]

Spec. Dr. Ayse Emel ÖNAL[a], Prof. Dr. Yakut Irmak ÖZDEN[a]
[a]*Istanbul University, Istanbul Medical Faculty, Departement of Public Health*

Abstract: Although an approach to take measures before problems actually occur is not possible in many developing countries, even for certain services that are regularly provided, it is hard but not impossible to devolp an understanding of service that is planned, implemented, and evaluated in advance of any unusual situation (i.e. disaster). This study aims to analyze the experiences gained in the 1999 Earthquake of Turkey's Marmara Region, and the studies concerning rehabilitation and preparedness for a new earthquake, which have been continuing from the earthquake until 2005, and to present acquired information from the point of view of public health, along with the experiences gained. On August 17, 1999, Turkey's Marmara Region was shook by an earthquake of a magnitude of 7.4 as measured on the Richter scale. The outcome was frightful: 17,480 deaths, 43,953 injyured, and 200,000 homeless. Futhermore, this earthquake gave rise to a multitude of serious public health problems such as difficulties in the provision of safe water, sewage problems, housing difficulties, those related to the collection of wastage, etc. The primary, as well as secondary and tertiary, health services systems were absolutely unprepared for such a disaster: There was no mobile hospital and no burn or trauma centers. The number of hospital beds and health personel were not sufficient enough to meet the needs. Intensive care, and hemodialysis services as well as the supply of blood were inadequate. Basic health[1] services could not be provided satisfactorily either. Due to the lack of personnel in the health centers, basic needs such as immunization, mother and child health services, the examination of patients, and laboratory analyses could not be met acceptabily. There were also difficulties in inhuming the deceased. In addition, the earthquake triggered a number of serious environmental risks. 6,500 tons of acrylonitrile ran out of the AKSA acrylic factory into the sea and the soil. A series of explosions happened at TUPRAŞ (petroleum factory) just after the earthquake caused very serious damage. Soon, after the earthquake came a period of rehabilitation and restoration activities such as settlement of homeless people in transitory homes, and accelerated construction of new houses and appartements for them; reopening of schools and workplaces; and providing psychological help and support to the survivors. In 2000, a National Earthquake Council was formed according to the European Code of Ethics Concerning Earthquake Prediction; And emergency preparedness plans have been processed by this council as well as by the local authorities, universities, and by various NGOs. Futhermore, many projects concerning disaster prevention and rescue are in progress. As a result, although it is hard to get prepared for an unusual situation such as earthquake, studies for preparedness for an earthquake have been accelerated in Turkey after the earthquake in Turkey's Marmara Region.

Keywords: Disaster, earthquake, preparedness, Marmara Region, Turkey

---

[1] The present work was supported by the Research Fund of Istanbul University. Project No. UDP-556/24062005.

Introduction

Disasters are natural environmental impacts that cause personal injury, physical damage, and massive economic collapse. According to the Stockholm declaration, disaster situations are characterized by "an acute and unforeseen imbalance between the capacity and resources of the medical profession and the needs of the victims or the people whose health is threatened, over a given period of time" [1]. In general, the results of unusual situations are worsened by the secondary factors. For instance, an earthquake may cause more death and damage if industrial facilities were established on inappropriate areas where no ground survey was performed and the buildings were constructed without consideration for the onset of an earthquake. It is a fact that earthquakes of the same magnitude can cause different results under different conditions and in different countries. Disaster situations tend to do more damage in poorly developed countries. Additionally, disasters give birth to a vicious circle worsening underdevelopment.

Each year, more than 1 million earthquakes happen in the world (approx. two each minute). Earthquakes caused approximately 1 million deaths in the last 20 years, with more than 80 percent being from eight countries including Turkey, China, Japan, Italy, Iran, Peru, former USSR, Chile, and Pakistan. The proportion of death/injured varies between 1:3 and 1:15 according to several studies. Injuries which require treatment constitute 3-40% of total injuries. Approximately 0.1-11% of injured people develops Crush Syndrome [2].

In recent years, the accelerated growth in urban population and the increase of population density have increased the vulnerability of such areas to deaths and injuries caused by earthquakes. An earthquake's magnitude, its proximity to a city place the issue of preparedness in extraordinary situations including whether or not prevention strategies exist and affect the damage and emergency situations to be caused by earthquake. To our knowledge, huge dams, underground nuclear tests and underground mining activities by digging deep wells can affect the aftermath of an earthquake. Landslides occuring after earthquakes, aftershocks, weather conditions, toxic substances released by affected industrial zones, and fires can also increase the earthquake's damages [2]. Although an approach to take measures before problems actually occur is not possible in many developing countries, it is hard but not impossible to develop an understanding of service that is planned, implemented, and evaluated in advance of any unusual situation (i.e. disaster).

Steps are required to prepare for the extraordinary situations including organizing the current humanitarian and technological sources in order to prevent or mitigate such extraordinary situations. Actions to be taken during or after any extraordinary situation should be organized in advance, so that everybody is prepared. "Extraordinary situation preparation plans" must be implemented during the situation as planned in advance (response). The response phase is followed by return, recovery, and treatment activities (recovery, rehabilitation) [3].

This study aims to search for the experiences in the 1999 earthquake that occured in Turkey's Marmara Region and studies concerning rehabilitation and preparedness for a new earthquake (prevention, mitigation and preparedness), which have been continuing since the earthquake until the year of 2005, and to present acquired information from point of view of public health.

1. Material and Methods

Internet keyword searches using the phrases "1999 Marmara Earthquake" and "Preparation Activities for Earthquake" were performed until 2005 in Turkey's Marmara Region—especially in İstanbul—in the Turkish language. An Internet scan was performed between May – July 2005. Information obtained from source Web sites were presented as findings, analyzing with public health approach in line with the writers' experiences of the Marmara Earthquake.

2. Results

An earthquake hit Turkey's Marmara Region on August 17th, 1999, with a Richter magnitude of 7.4; its epicenter, the city of İzmit (Kocaeli). As a result, 17,480 people died, 43,953 people were injured, and 200,000 people lost their homes [4; 5]. According to the experiences of earthquake survivors, seawater was warmer for one week compared to before the earthquake, there were waves the height of 1.5 meters, and a mist layer gathered above the sea (radon gas) [5; 6]. The city of Kocaeli, with a population of 1,177,379 and its district Gölcük, the city of Sakarya (Adapazarı) with a population of 731,800; the city of Yalova, with a population of 163,916, and some districts of İstanbul, namely, Avcılar with a population of 214,621; Bağcılar with a population of 487,869; and Küçükçekmece with a population of 456,463 were the most damaged. This region has an industrial production capacity reaching 52.7% of the total of Turkey's goods (food, petrochemicals, glass, automotive, textile, clothing, electrical goods, ships and railway car). Approximately 40% of workers employed in the entire country of Turkey are located in this region [7].

Subjects of high priority during the earthquake were: search and rescue activities, first aid activities, emergency medical aid, urgent communication and transportation. Turkish Armed Forces, the Ministry of Health, the Ministry of Public Works and Settlement, the General Directorate for Disaster, the General Directorate for Civil Defense, governorships, municipalities, the Turkish Red Crescent, professional establishments, NGOs, and other international institutions took an active role during the earthquake. (Response Period in the Earthquake) [3]. Turkish Medical Association established an Extraordinary Situations Working Unit. Turkish Psychiatry Association established a Disaster Psychiatry and Mental Health Working Unit [8].

On August 17, 1999, the Ministry of Health created a Crisis Center that worked on 24-hours per day. Center's duties included: to plan and evaluate the health services within the Region; to eliminate difficulties occurred; to evaluate the orders from Primeministry Crisis Center; to meet medical equipment, personnel, tools and other needs, and; to provide supplies received from Turkey and abroad to the disaster region in a planned manner. Crisis centers were created in provincial health offices. Since then, they have been re-structured as Main Contagious Diseases, Environmental Health, Emergency Aid and Rescue Services, Personnel Office, Administrative and Financial Office [7].

The death-injured ratio in Marmara Earthquake was 1 to 2.5. A total of 5,302 patients who were placed in 35 Reference Hospital after Marmara Earthquake represented 12 percent of total injured people. Acute renal problems were found in

639 patients out of 5,302. 477 of them required dialysis support. Patients requiring dialysis were 1.1 % of total injured people [9].

Ambulance and first aid services were insufficient. According to data given by the Ministry of Health, 113 ambulances (34 to Kocaeli, 58 to Sakarya, 21 to Yalova, 10 to Bolu), 154 medical specialists, 116 medical practitioners, 197 nurses, and 80 medical personnel were sent by the 112 Emergency Search and Rescue Unit [7]. However, problems were faced in the course of providing health services due to building, infrastucture problems and insufficient personnel. Bed capacity of the hospitals was also insufficient. According to the determinations made after the earthquake, the bed capacity in Kocaeli and Adapazari was decreased by 28% and 33%, respectively, compared to the period before the earthquake [10]. 20,868 injured people were transported to other medical institutions in surrounding cities after the earthquake [7]. 12,490 of them received outpatient treatment, 8,378 inpatients, 1,881 underwent surgery, and 708 were transported to other hospitals [7]; problems also occured during burials; formaldheid was insufficient [10].

There were no field hospitals; "Tent Policlinics" were established. Loyal medical experts and medical practitioners gathering from neighbor cities worked there. Intensive care, hemodialisis, and blood bank services became insufficient. There were no burn or trauma units [10].

After the first five days when emergency cases had been almost settled, activities for basic health services that had been initiated in the region from the first day of the earthquake were continued. Public health laboratories were created in Adapazarı Toyota Sa Hospital, İzmit State Hospital, and Yalova State Hospital [7]. Water and food were analyzed. Activities for disinfection and fight against vectors were initiated. Tetanus vaccine and serum, gas gangrene serum, rabies vaccine and serum, scorpion serum and snake serum were immediately despatched to the earthquake region. The Ministry rapidly met the vaccine needs (i.e., diphteria, pertussis, tetanus, polio (DBT-P), measles, Hepatitis B, BCG, and DT) arising from failure in the cold chain due to blackout. All kids under 14-years of age were vaccinated against measles. No contagious disease epidemic occured although potential typhoid fever, paratyphoid fever, dysentery, other gastrointeritis, and Hepatitis A epidemics were expected. People were informed of health rules to be obeyed after the earthquake by means of verbal and written announcements. Although the presence of such commenced activities, certain difficulties occured while basic health services were provided. Due to lack of personnel in village clinics in Adapazarı, certain basic needs such as immunization, family planning, parturition services, medical examination, and laboratory services could not be met. There were no physicians in two village clinics in Gölcük [10]. Basic services, such services as clean water, sewer system, accomodation, collecting garbages could not be fulfilled adequately. Drinking water was derived from private companies, municipalities, city grid with limitation, drills, and wells. At the end of the first year following the earthquake, 33% of the population in Gölcük was yet to be supplied with clean water. 70% of the population in Sakarya was unable to benefit from sewer system services [10]. (Normally, 87% of people living in rural areas and 96% of those in cities are able to access clean water in Turkey, while the proportion of those who are provided with sewer system service is 55% [11; 12].

There were people who were still living in tents in Kocaeli even after one year following the earthquake [10]. Following the acute stage in Marmara Earthquake, the health problems faced in later stages were mainly pyschological problems. Around 500 people from Adapazarı region were examined between eight and 11 months for pyschological problems related to the earthquake. It was determined that 70% of this group who did not report any obvious pyschological problem was actually in need of pyshosocial support [8]. In addition to such a traumatic experience, limited accomodation, education, and job opportunities caused the psychiatric problems to continue long after the earthquake. Pyschologic rehabilitation was insufficient despite the efforts of the Ministry of Health, NGOs, professional associations, and other international institutions.

After the earthquake the explosion in TÜPRAŞ petroleum filling facility and 6,500 tons of acrylonitrile from AKSA acrilic industry factory having dispersed in water and soil added to the impacts of the earthquake; certain negative environmental health effects became much greater. TÜPRAŞ fire continued for three days. Sea animals died because of the pollution. Animals and plants died [13; 14]. It is known that acrylonitrile outflowing from AKSA causes acute lung edema and sudden death. Six people from the region died suddenly and suspiciously. People affected by acrylonitrile indicated following symptoms: Itching and inflammation of the skin, watery eyes and inflammation of the eyes, fainting, loss of balance, inflammation and ache in throat and nose, hoarseness, irritation in respiratory tract, stertorous respiration, dry hacking cough, blue lips, stomachache, nausea, vomiting, and bloody diarrhea. Chronic effects to be observed in the long term after contact with acrylonitrile were hair and weight loss.

According to the literature, carcinogenic effect, teratogenic effect, mutagenic effect, thyroid gland and endocrine system illnesses and illnesses relating to the loss of immunity system are expected as late symptoms of acrylonitrile [15].

A recovery period began after the earthquake. Transporting homeless people from temporary housing (tent cities and prefabricated houses) to permanent houses, opening the schools, reliving the business life, providing pyschologic support and rehabilitation for earthquake survivors were among them. There were numerous projects which were initiated and carried out under this scope. Urgent Aid Project for Minor Farmers Effected from Marmara Earthquake TCP/TUR/0065 supported by FAO is one of them [16]. With the support of Ministry of National Education (MEB), Social Services Child Protection Institution (SHÇEK) and UNICEF, the academic year was started in tents and schools [17]. The Ministry of National Education and UNICEF initiated psychologic support programs in schools. [18; 19]. Credits obtained from international financial institutions were utilized for constructing buildings, infrastructure and sewer system grid, new hospitals, and rehabilitation of health units, primary schools, and rehabilitating transportation systems with the railroads, at first, supporting minor and medium level enterprises. Credits were obtained from Japan, Spain, South Korea, Belgium, IMF, World Bank, Gulf Cooperation Council, Islam Development Bank, European Council Development Bank, European Investment Bank [20].

According to releases made by the Prime Minister Bülent Ecevit in 2001, 44,000 prefabric houses had been built in the 1999-2001 period, and 27,000 houses out of 40,000 permanent houses under construction had been completed. 117 schools and 947 additional class rooms were under construction. Sixteen prefabric health facilities with a capacity of 1,000 beds were established by the Ministry of Health and Social Security Institution. Two permanent hospitals were constructed with a total of 220 beds, each. Agriculture Bank (Ziraat Bankası) provided 4.5 trillion credits for 3,000 farmers. Mandatory Earthquake Insurance was implemented in Turkey for the first time [21].

As understood from the explanations of the Prime Minister Bülent Ecevit, prevention, mitigation, and preparedness activities were initiated and continued for the next potential earthquake, in addition to recovery and rehabilitation activities. In 2000, an autonomous National Earthquake Council was established under European Code of Ethics Concerning Earthquake Prediction [22]. A Turkish Emergency Management General Directorate was formed in the Prime Ministry [21].

Numerous projects and studies aiming at disaster prevention-mitigation and rescue have been commenced. One of them is Marmara Earthquake Emergency Reconstruction Project, which was funded by credit obtained from World Bank. It was concluded in the year of 2004 [23]. Another one is Microzonation and Disaster Prevention/Migration Project carried out by Istanbul Metropolitan Municipality (IMM) and Japan International Cooperation Agency (JICA). It was concluded in July 2003 [24].

Activities carried out by Istanbul Metropolitan Municipality with respect to early intervention and emergency warning system included: Earthquake Master Plan under Istanbul Metropolitan Municipality Disaster Coordination Center (AKOM), which is a joint project with Istanbul Technical University (ITU), Middle East Technical University (METU), Yıldız Technical University (YTU) and Boğaziçi University (BU), Zeytinburnu Urban Regeneration Project, Urban gjeology and activities for producing land utilization maps, establishing Earthquake stations (TÜBITAK, Boğaziçi University Kandilli Observatory and with the cooperation of Japan), making earthquake predictions (with the cooperation of TÜBITAK) such as physical and chemical changes in spring waters and daily monitoring of radon gas, Earthquake Park Project, Disaster Station for each Neighborhood Project [25].

Earthquake Emegency Action Plans were prepared in governorships, universities, municipalities, health institutions, and many public institutions in order to determine what would be done during and after another earthquake. Plans by Beşiktaş District Municipality of Istanbul, Üsküdar District Governorship of Istanbul and Turkish Red Crescent can be given as an example [26; 27; 28]. Public awareness campaigns are included in disaster activities by almost all institutions mentioned above. Boğaziçi University Kandilli Observatory and Earthwuake Research Institute commenced a Training Program of Preparedness against Earthquake. One can access to Family Preparedness Plan against Earthquake through the Internet [29]. A project under the name of "Basic Disaster Awareness Project for Turkey" was initiated by the Ministry of National Education and Boğaziçi University Kandilli Observatory Earthquake Research Institute; the U.S. Agency for International Development and the Office of Foreign Disaster Assistance are supporting this project [30].

According to Boğaziçi University Kandilli Observatory Earthquake Research Institute, 10 early warning systems were established in Marmara for any possible earthquake in Istanbul. Shockwave will be determined 22 seconds before. The first signal will reach Kandilli in 13 seconds. Natural gas and power lines will be cut within the remaining 9 seconds [31].

Istanbul Metropolitan Municipality (IMM) planned the services to be provided in case of emergency as search-rescue and firefighting, temporary settlement services, medical services, transporation services, damage determination activities, logistics coordination services, food and drink services, debris removal services and agricultural services [32].

The Ministry of Health established a Department of Health Services in Disasters for preparation activities against earthquake and other disasters in 2004 under Presidency of Research, Planning and Coordination Board. In 2005, it was decided to establish Health Services Unit in Disasters in order to directly serve as Health Officers in Provincial Health Offices [33]. In the first stage, National Medical Research Squads (UMKE) were established in 11 provinces (Adana, Afyon, Ankara, Bursa, Diyarbakır, Erzurum, İstanbul, İzmir, Samsun, Sakarya, Van) and in Kocaeli, Yalova and Düzce provinces, where the earthquake risk is high and squad members were trained [34].

3. Conclusions

In conclusion, Turkey, as a country located on Alpen-Himalayan seismic zone where earthquakes frequently happen, is under eatrhquake risk with various magnitudes [35]. The Marmara Earthquake that occured on August 17, 1999, was a difficult experience for Turkish people. Certain lessons were learned from these experiences:

1. Turkey was not ready for a big earthquake.
2. Buildings must be enforced against earthquakes. New settlement areas must be built by conducting ground surveys. Industrial zones and other settlement areas should not be built on fault lines. Buildings must be earthquake-proof. Earthquake insurances must be ensured.
3. Early warning systems must be established. So, at least such precautions as cutting power and natural gas lines, rapid evacuations in industrial zones and settlement areas can be taken.
4. Everyone and every institution should have an emergency action plan concerning what will be done in case of earthquake.
5. The first five days following an earthquake are critical in terms of search and rescue, first aid and emergency medical attention. Rescue activities must continue for a minimum of five days after the disaster. It must be planned how to insure transportation and communication within this period.
6. Hospitals must be prepared with the capacity to meet all applications within one week.
7. Intensive care units, hemodialisis units, blood banks, burn and trauma units must be established according to the necessity and the capacity of those existing must be increased.
8. All other medical institutions must have necessary hardware in proportion to the population in terms of infrastructure, medical equipment, tools and personnel needs.
9. Medical training for disaster should be continously given to medical personnel who are living in areas under earthquake risk.
10. The public must be kept aware of earthquake risks by performing continuous activities.

REFERENCES

[1] Saçaklıoğlu F, Sarıkaya Ö. 2002. Basic concepts concerning extraordinary situations. Turkish Medical Association, Medical Services in Extraordinary Situations, in Handbook for Medical Personnel, Ed: Karababa AO, 2nd Edition, Ankara;. 11-20p.

[2] Vatansever K, Türk M, Vatansever M. 2002. Epidemiologic Characters of Extraordinary Situations. Turkish Medical Association, Medical Services in Extraordinary Situations, in Handbook for Medical Personnel, Ed: Karababa AO, 2nd Edition, Ankara; 21-40p.
[3] Durak Hİ, Vatansever K, Karabilgin ÖS. 2002. Preparedness in extraordinary situations. Turkish Medical Association, Medical Services in Extraordinary Situations, in Handbook for Medical Personnel, Ed: Karababa AO, 2nd Edition, Ankara; 203-222p.
[4] 2000 August 18. Radikal Newspaper. Available from: http://www.belgenet.com/deprem/170800_r18.html
[5] 2000 August 17. Cumhuriyet Newspaper. Available from: http://www.belgenet.com/deprem/170800_c17.html
[6] E Atilgan, Bitter facts of 17 August Earthquake.Available from: http://www.istabip.org.tr/hf/2000-8.asp
[7] Aycan S. 1999. Marmara Earthquake and Services given by the Ministry of Health. Available from: http://www.spgk.saglik.gov.tr/tr/afet_yonetim/sunum1.htm
[8] Yüksel Ş, Phsychological problems in earthquake zone. Available from: http://www.ttb.org.tr/TD66/7.html
[9] Sever MŞ. Crush syndrom and what the Marmara earthquake taught. Available from: http://www.tsn.org.tr/egcalhek/crush_sendromu.pdf#search='crush%20sendromu%20deprem'
[10] 1st year report for Marmara earthquake prepared by Turkish Medical Association. Available from: http://www.istabip.org.tr/hf/2000-8.asp
[11] UNICEF 2005. The State of the World's Children
[12] Demirer GN, Abay T. 2002. Environment in Turkey and the world in the year of 2020, what kind of world between 2000-2020, Turkey, Health, Medical Environment may be envisioned and succeeded, Turkish Medical Association, Ankara, 2002.
[13] 17.11.1999-17 August-AKSA accident/Precautions that must be taken in terms of health/lessons learned. Available from: http://www.istabip.org.tr/yk/ykbaarsiv1999.asp#15
[14] Tüpraş, Faith or Expected Result? Available from: http://www.ato.org.tr/dergi/1999_3/gundem1.html
[15] Bakırcı N. An industrial environment disaster: acrylonitrile. Available from: http://www.ttb.org.tr/isak/msg/ocak01/9.htm
[16] Available from: http://www.un.org.tr/fao/faal_afet.htm
[17] Available from: http://www.unicef.org/turkey/pc/_ep9.html
[18] Available from: http://www.unicef.org/turkey/sy8/_ep13.html
[19] Available from:http://www.meb.gov.tr/Stats/ist2000/b9a.htm
[20] T.C. Prime Ministry Undersecretary of Treasury, Consultancy of Press and Public Relations, Press Release. Available from: http://www.treasury.gov.tr/duyuru/basin/DepremBilgi160801.htm
[21] Releases by Prime Minister Bülent Ecevit. Available from: http://www.belgenet.com/deprem/17082001_02.html
[22] Available from: http://udk.tubitak.gov.tr/
[23] National Earthquake Council.Available from: http://www.worldbank.org.tr/WBSITE/EXTERNAL/COUNTRIES/ECAEXT/TURKEY
[24] Projects and studies aiming prevention-mitigation and rescue in case of a disaster in Istanbul. Available from: http://aym.istanbul.gov.tr/proje.htm
[25] Activities by Istanbul Metropolitan Municipality Disaster Coordination Center (AKOM). Available from: http://istanbul.bel.tr/deprem/deprem/calişmalari.htm
[26] Available from: http://besiktas.beltr/channels/1.asp?id=210
[27] Beşiktaş municipality disaster plan, Üsküdar District Governorship, Disaster Management Center. Available from: http://www.uskudar.gov.tr/afetyonetim/afetyonetim.htm
[28] Available from: http://www.kizilay.org.tr/channels/1.asp?id=27
[29] Kandilli Observatory and Earthquake Research Institute, Training Program for Preparedness against Disaster.
[30] Available from: http://www.meb.gov.tr/duyurular/duyurular/TemelAfetEgitimiProjesi/TemelAfetEgitimiProjesi.htm
[31] Available from: Istanbul.com.tr/istanbultr/350/35001/kent_haber_detay.cfm?sayfa=6848
[32] Available from: http://aym.istanbul.gov.tr/
[33] Available from:http://www.saglik.gov.tr/sb/extras/mevzuat/asopgen.doc
[34] http://www.saglik.gov.tr/sb/extras/mevzuat/2005UMKEGTGEN.doc
[35] http://www.deprem.gov.tr/depobölge/index.htm

Scientific Networking and the Global Health Network Supercourse
F. Linkov and R. LaPorte (Eds.)
IOS Press, 2006

© 2006 IOS Press. All rights reserved.

Impact of Environmental Factors on Psychic Disorders and Somatic Diseases in Lithuania

Saulius SLIAUPA[1,a], Gediminas ZUKAUSKAS[b], Larisa KORABLIOVA[c], Zydrunas DENAS[d], Algimantas ZAKAREVICIUS[e], Marija JAKUBENIENE[b] and Ona DAVIDONIENE[f]

[a]*Institute of Geology and Geography, Lithuania*
[b]*Institute of Forensic Medicine, Lithuania*
[c]*Lithuanian Geological Survey, Lithuania*
[d]*Vilnius University, Department of Geology and Mineralogy, Lithuania*
[e]*Vilnius Gediminas Technical University, Lithuania*
[f]*State Mental Health Centre, Lithuania*

Abstract. Human health is influenced by a complex interaction of different factors, including the environmental parameters. The impact of anomalous rates of different geo-factors is well known (e.g., natural and man-made geochemical anomalies), whereas the influence of the background rates of environmental parameters receives little acknowledgement. This paper discusses the statistical relationship of the different environmental factors to selected human diseases on a countrywide-scale in Lithuania. Five groups of geo-factors are analyzed: (1) gravity and magnetic fields, (2) concentrations of geochemical elements in the soil, (3) sub-soil lithologies, (4) relief and some landscape features, (5) drinking water chemistry. Environmental factors were correlated with the spatial variations of the schizophrenia, alcoholic psychosis, suicides, cancer, infection, cardiovascular, respiratory diseases. The statistical approach was applied to ensure an objective identification of the interacting factors. The different formats of data were unified using GIS techniques that enabled scaling of geological databases to 45 administrative regions of Lithuania. Some strong correlations were identified that imply the importance of the living environment on the human health and psychic disorders.

Keywords: environmental health, geomedicine, geo-factors, psychic health, suicides, GIS

Introduction

Human health is influenced by the interaction of factors with different factors including the environment [1; 2; 3]. Geomedicine has roots that go back centuries; papers appeared in the 19th century noting geographical patterns in the distribution of diseases. Selected geological factors were studied with respect to the human health in Lithuania. Five groups of geo-factors were analyzed: (1) gravity and magnetic fields,

[1] Corresponding Author: Saulius Sliaupa, Institute of Geology and Geography, T. Sevcenkos 13, LT 03123, Vilnius, Lithuania; E-mail: sliaupa@geo.lt.

(2) sub-soil lithologies, (3) concentrations of chemical elements in the soil, (4) relief and other landscape features, (5) drinking water chemistry. Environmental factors were correlated with the spatial variations of the (i) psychic diseases and human behavior such as the schizophrenia, alcoholic psychosis, suicides, homicides and (ii) somatic diseases such as the tuberculosis, cancer, infections, cardiovascular, and respiratory diseases.

It is conventionally realized that chemical elements have an impact on human health. The distribution of trace elements in the soil is varies across Lithuania, depending upon human induced pollution and natural geochemical load related to the soil and sub-soil composition, vegetation, relief, etc. The deficiency or excess of chemical elements in the living environment can negatively influence human health, whereas other elements show a positive impact. Also, the landscape may play a certain role, e.g., as the psychological environmental factor. The most controversial is the influence of the earth's potential fields [4; 5]. Geomedicine studies mainly concern the influence of the geological environment on the human health, whereas human behavior is less considered. It is recognized that people exposed to various natural disasters (e.g., floods, earthquakes) could develop psychological sequelae from stress such as major depression, chronic anxiety [6].

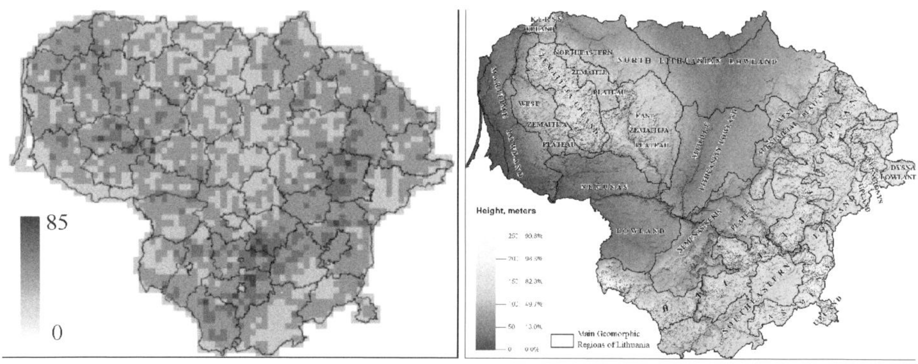

Figure 1. Comparison of density of settlements (grid 10 km) with relief, Lithuania. The high-topography dissected areas show higher density than the low-topography landscape.

Lithuania is characterized by a rather low intensity of geological processes and relatively miserable variations of environmental parameters. Still, the living environment is believed to have a certain impact on human health and behavior. The knowledge and recognition of those relationships is an important issue for developing population control [7]. The rapidly increasing globalization and mobility of the present-day society in Lithuania seems to obliterate variations induced by environmental parameters and are often underestimated. A comparison of the relief (in turn, it is related to other parameters, such as sub-soil lithology, geochemical enrichment of soil, land cover, etc.) and density of settlements provide a spectacular illustration of the considerable impact of the landscape on the social framework (see Figure 1).

1. Data and Methods

Statistical data on distribution of the schizophrenia, alcoholic psychosis, suicides, tuberculosis, cancer, infection, cardiovascular, and respiratory diseases were collected from 45 administrative regions of Lithuania in 1999 [8; 9]. The data was considered separately by male, female, and average disease rates. The more profound data are available for the cardiovascular diseases and suicides covering the period 1999-2003. A comparison of these data indicates the strong persistence of major trends that imply involvement of particular factors (Figure 1).

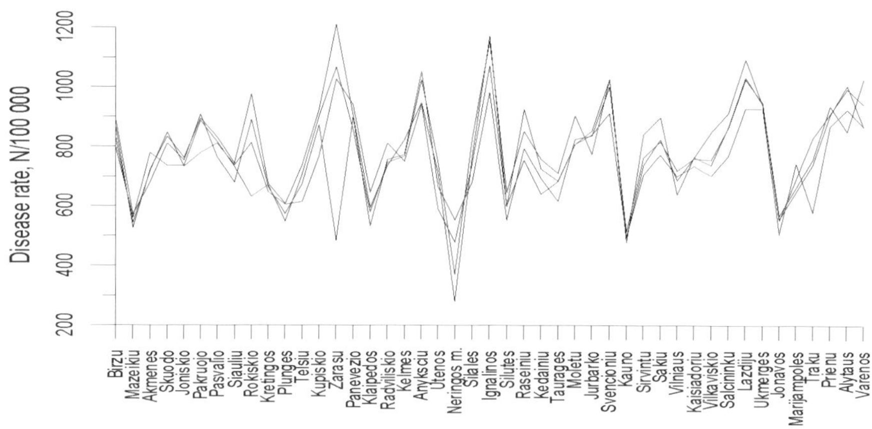

Figure 2. Cardiovascular disease rates (lethal events per 100,000 population) in different regions of Lithuania, year 1999, 2001, 2002, 2003. Note persistence of trends.

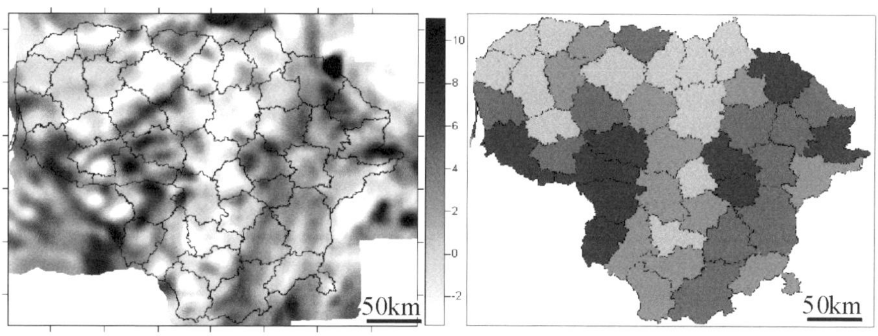

Figure 3. Conventional magnetic field map (left figure) scaled to 45 administrative regions of Lithuania (right figure).

As conventional maps of the environmental factors are designed in a different way (contour lines, polygons, polylines, points), a special technique was applied to unify geo-factors to separate administrative regions (Figure 3). The present study is based on application of the statistical methods and geographic information system (GIS) tools seeking to identify correlations between parameters of different formats. GIS technology makes it possible to link or integrate information that is difficult to associate any other way [10]. GIS is capable of performing a wide-range of information processing. The average rates, variability, dispersion values of geo-parameters were calculated (and plotted) using GIS tools [11]. Data were processed and interpreted in MapInfo Professional environment. For automation of tasks, study-specific programs were written in MapBasic language. The statistical correlation between administrative region-scaled environmental and health parameters were obtained using STATGRAPHICS Plus program. Only correlations with confidence level 0.95 and correlation coefficient higher than ±0.30 were considered to be consistent.

Digital models of the gravity and magnetic fields of Lithuania [12] were averaged to administrative regions. The relief height, slope gradients, dissection were derived from the digital terrain model of Lithuania [13]. The other analyzed landscape factors are the percentage of forest, lakes, rivers, and wetlands that were derived from the CORINE Land Cover model. The density of the neotectonic faults and occurrence of the hydrogeochemical anomalies related to active faults [14] were incorporated into analysis. The sub-soil lithologies were derived from the digital Quaternary Map of Lithuania [15]. They show different geochemical enrichment of the soil. Twenty-eight trace elements [16] were analyzed. The groundwater is mainly exploited for drinking water supply in Lithuania. Data on the chemical composition of the drinking water were collected from more than 12,000 boreholes stored in the GEOLIS database of the Lithuanian Geological Survey.

2. Disease Rate Distributions

Similarly to other regions that show geographic dependence of the medical parameters, [17; 18; 19] the psychic disorder and human health parameters show rather distinct trends across Lithuania, implying involvement of the geographic-depended factors. In particular, the northeastern part of Lithuania is characterized by a higher schizophrenia rate. In the west, the distinct NW-SE trends are identified, whereas the eastern half of the country is dominated by NE-SW anomaly pattern. The highest disease rate of the alcoholic psychosis is registered in the west; it is the lowest in central Lithuania. Central Lithuania is marked by increased suicide rate. It should be noted that Lithuania has the highest suicide rate in Europe [20]. Also, the rate of tuberculosis is quite dramatic, averaging 280 events per 100,000. The highest rates are reported from west Lithuania, and the distinct trend crossing the whole country from northwest to southeast. In general, the eastern part of Lithuania is characterized by lower tuberculosis rate. Cancer shows a different pattern. The highest disease rate is registered in southwest Lithuania, while the lowest is in some regions of east Lithuania. The rate of cardiovascular diseases is generally higher in the eastern part of Lithuania. East Lithuania is characterized by higher respiratory diseases risk, and is essentially high in the south. The increased infection rates are reported from central and southwest Lithuania.

3. Results

The analyzed somatic diseases and mental disorders show strong statistical correlation with some environmental factors. Cancer has a consistent correlation with the magnetic field, essentially with its variability (correlation coefficient +0.45) and some geochemical elements of the soil, i.e. the titanium and zircon (+0.32), and some weaker relationship to Y, Yb, Nb (+0.22 to +0.29). Tuberculosis reversely correlates with the gravity and magnetic fields (-0.33 and -0.35). No consistent relationships are identified for the soil trace elements, whereas, drinking water containing increased amount of Ca and HCO_3 statistically has the remedying effect.

Statistically consistent relationship of the respiratory diseases is identified with the differentiation of the relief (+0.44), essentially in the areas that show less dense river network (correlation is -0.32). Of the analyzed soil chemical elements, Rb and Sr have a consistent negative loading on the respiratory disease rate, as well as the drinking water salinity and SO_4 content (-0,38).

Infectious diseases well correlate with the gravity and magnetic fields, the correlation coefficients are respectively +0.31 and -0.33. No evident correlation is identified for the other environmental parameters.

Cardiovascular diseases correlate with the magnetic field (+0.37). The other directly correlating factor is the density of the lake system. The low disease rate is related to higher concentrations of Na and K, higher pH, lower amount of iron (+0.41) in the drinking water.

Schizophrenia shows the direct correlation with the magnetic field (+0.30), whereas soil trace elements P, Pb, and Sn correlate reversely (-0.40, -0.38, -0.40). These elements compose a strong association showing a similar pattern across the country, their concentrations increase in the sandy soil. The schizophrenia rate also shows some decrease of about 30 percent with increasing concentration of Li in the soil. Of sub-soil lithologies analyzed, the Aeolian and glaciolacustrine sedimentation areas show the strongest reverse correlation with the schizophrenia. The direct correlation was identified for the geochemical enrichment index of the pit (+0.39). Ca, Mg, HCO_3, SO_4, Fe concentrations in the drinking water and its hardness directly correlate with schizophrenia, whereas the increased concentrations of Na, K, Cl decrease the disease rate.

Alcoholic psychosis shows strong correlation with some landscape features, such as the direct relationship to the relief height and dissection. The soil enriched in Yb and Y (and some weaker indications for La, Li, Sr) has statistically lower disease risk. The water chemistry is an essential parameter, as shows the reverse loading of Mg, Ca on the alcoholic psychosis.

The suicide level also has a clear geographical control. It reversely correlates with the magnetic field, relief dissection, forest, and lake percentage. The areas dominated by sandy lithologies show statistically lower suicide level, essentially the glaciolacustrine basins, whereas the bottom morainic lowlands coincide with increased suicide rates; water chemistry also well correlates with the suicide level, i.e., Na, K, high pH showing "remedying" effect.

Homicide rates do not show any correlation with environmental factors, except the fault density that positively correlates with the homicide rate.

4. Discussion

Human health is shown to correlate with environmental factors. However, statistical correlation obtained in this study does not necessarily imply the genetic relationship, as environmental parameters are intimately interrelated. For instance, it is hard to believe that the percentage of lakes has anything to do with the cardiovascular diseases, or that relief might have any direct impact on the suicide rate. Rather, they reflect some particular environmental systems with a specific set of factors, such as the soil lithology and chemical composition, drinking water chemistry, etc.

The statistical cross-correlation of the environmental parameters reveals two large groups that can be conventionally called the gravity and magnetic field groups. The present relief and sub-soil lihtologies were formed by the Quaternary glaciations and de-glaciations processes. In turn, these processes were strongly controlled by the deep structure of the earth's crust that is sensitively reflected by the magnetic and gravity fields. The former reveals the distribution of the shallow geological bodies, whereas, the latter one is influenced mainly by deeper geological sources. Accordingly, the gravity and magnetic fields relate reversely in terms of statistical correlation. It was defined that the intensity and variability of the magnetic field has a direct statistical correlation with the present relief (height, dissection), whereas the gravity field has a reverse statistical load on the relief. Furthermore, the sandy sub-soil lithologies are more common for high magnetic field areas, whereas the high gravity field directly correlates with clayey lithologies. The relief and related lithologies are the major factors controlling the other landscape parameters (drainage network, land cover, etc.). Accordingly, the higher magnetic field (in turn, the higher and more dissected relief) directly correlates with the lake, wetland, forests percentages, while reversely correlating with the river and neo-tectonic fault network densities. These factors show opposite trends with the gravity field. The relief intimately relates with the sub-soil lithologies that have different geochemical load.

The statistical cross-correlations between the environmental factors explain obtained statistical relationships with diseases. Cancer is shown to correlate with the magnetic field and some trace elements (Ti, Zr, Y, Yb). These elements are typical representatives of the clastic (allochtonous) association showing increased amounts in sandy lithologies. This particular lithotype set is believed to be the environmental factor directly controlling the spatial distribution of the disease rate in a country scale.

As for the respiratory diseases relief is of the primary importance showing the higher rate in the high-dissected hilly areas. Atmospheric flows and related chemical enrichment are strongly dependent on the relief that can explain the obtained relationship. Also, these areas (central and west Lithuania) are characterized by some enrichment in SO_4. It is difficult to predict the extent the drinking water can influence respiratory diseases. As for tuberculosis, pH is of primary importance among the environmental factors, as indicated by a strong reverse correlation with the soil and water pH (respectively, drinking water enriched in Ca and HCO_3 has a positive effect on the health).

Infections are the "gravity" group controlled (direct correlation), though they do not show any strong statistical correlation with any particular factor, except soil pH. The more acid soil seems to have remedying effect on infections.

By contrast, cardiovascular diseases directly correlate with the "magnetic" group factors, but the most essential parameter seems to be the drinking water chemistry, the lower disease rate associating with high concentrations Na, K. These two elements, especially potassium, are well recognized factors helping to prevent such diseases. The influence of the water chemical composition on the heart function is well established in the other regions [21; 22; 23]. Soil impact can be also important [24].

Schizophrenia is the "magnetic" group related. The major parameter is likely the sandy litholgies that are typically enriched in phosphorus, which is known to have strong impact on brain function [25; 26]. The strong correlation of the schizophrenia with the geochemical enrichment of the pit was obtained. The high- and low-peatlands show different geochemical associations, such as much higher concentrations of most trace elements in the low-peatlands. The type of the peat is very sensitive to geological-geographical conditions. Low-peat is common in the low-topography areas dominated by glaciolacustrine and Aeolian sub-soil lithologies, also in some high-topography areas covered by the same type of sediments. High-peat is preferentially distributed in the high-topography till-composed areas. The concentration of phosphorus is accordingly higher in the former environment dominated by sandy lithologies. Also, schizophrenia correlates with drinking water chemistry. It is difficult to judge if it reflects the genetic relationship. Yet, potassium is known to have a remedying impact and, therefore, the water can be also important for the disease rate variations.

Alcoholic psychosis shows the strongest correlation with drinking water, though other factors ("magnetic" group) should not be neglected (e.g., correlation with lithium), too. The increase in Mg concentration for about two times results in reducing the alcohol psychosis rate for four-five times that is an expected effect, as an excessive consumption of the alcohol leads to deficiency in Mg in an organism, and water chemistry can either promote or inhibit this process.

Suicide level shows surprisingly strong geographical control. A likely explanation is an impact of the landscape on the economic-social framework. On the other hand, direct involvement of the geo-factors should not be neglected; as it is hinted by the strong correlation of the disease rate with the drinking water chemistry, in particular the remedying (in terms of statistics) effect of potassium. It is known that when the potassium starvation becomes chronic in proportions, suicide is often contemplated and sometimes attempted. It should be noted that suicide rate does not show any correlation with the alcohol psychosis (it is even slightly reverse).

The homicide rate does not show any correlation with environmental factors, as it is primarily the social-dependent parameter. Still, some interesting direct correlation of the fault density with the homicide rate should be mentioned. It is often discussed to what extent tectonics may have an impact on crime rate. Here we can state that at least obtained statistical relationships are in favor of this hypothesis.

5. Conclusions

Environmental variations are rather minor in Lithuania. However, even minor changes in the living environment seem to have a certain impact on the human health. We fully acknowledge that human health depends on interaction of very different factors, and the environment is one (and maybe not the most important) of those factors. Still, the obtained consistent statistical correlations leave no place for doubts that geo-factors should be taken into account.

References

[1] Cannon HL, Hopps HC. eds. 1970. Geochemical environment in relation to health and disease. Geological Society of America Special Paper; 140:289p.
[2] Draggan S, Cohrssen JJ, Morrison RE. eds.1987 Environmental impacts on human health. New York: Praeger.
[3] Bunnell JE. 2004. Medical geology: Emerging discipline on the ecosystem-human health interface. EcoHealth 1:5-18.
[4] Dubrov AP. 1998. Geomagnetic field as the major synchronizer of biorhythms. Parapsychology and Psychophysics 1(25), Moscow, 30-32. (in Russian).
[5] Melnikov EK, Rudnik VA. 2000. Analysis of impact of the low-intensity geophysical and geochemical anomalies related to active faults on health of population of St. Petersburg region. Low- and very-low-intensity fields and radiation in biology and medicine. International Congress, St. Petersburg, 3-7.07.2000, 90-92. (in Russian).

[6] Baum A, Fleming I, Israel A, O'Keeffe MK. 1992. Symptoms of chronic stress following a natural disaster and discovery of a human-made hazard. Environ Behav 24:347-367.
[7] Remington PL, Smith MY, Williamson DF. 1988. Design, characteristics, and usefulness of state-based Behavioral Risk Factor Surveillance: 1981–87. Public Health Rep 103:366–375.
[8] NOMESCO 2001. Nordic/Baltic health statistics. Nordic Medico Statistical Committee 62:95p.
[9] Statistics of Lithuania. 2001. Causes of death. Vilnius; 264p.
[10] Sliaupa S, Belickas J, Denas Z. 2003. Integrated approach and application of GIS for management of geological data. Technika Poszukiwan Geologichnych. Geosinoptika i Geotermia 6:63-72.
[11] Korabliova L, Popov M. 1998. Digital models of gravity and magnetic fields of Lithuania, scale 1:200 000. Geological Survey of Lithuania: Annual Report for 1997, Vilnius, 21-23.
[12] Denas Z, Kumetaitis A, Sliaupa S, Zakarevicius A. (2005). Application of GIS for surface erosion model of Lithuania. Social Strategies 40. Peter Lag AG.
[13] Sliaupa S. 2000. Neotectonics as a hazardous factor. Geological Survey of Lithuania: Annual Report for 1999, Vilnius, 35-37.
[14] Guobyte R. 1999. Quaternary geological map of Lithuania. Geologijos Akiraciai, 1/99. Vilnius, 7–16.
[15] Kadunas V, Budavicius G, Gregorauskiene V, Katinas V, Kliaugiene V, Radzevicius A, Taraskevicius R. 1999. Geochemical Atlas of Lithaunia, Vilnius, 91 p.
[16] Sauer HI, Brand FR. 1971. Geographic patterns in the risk of dying. Environmental geochemistry in health and disease, edited by H.L. Cannon & H.C. Hopps, Geological Society of America Memoir 123:131-150.
[17] Villanueva W. 1982 September. The cartography of death. Discover 82-85p.
[18] Pickle LW, Su Y. 2002 . Within-state geographic patterns of health insurance coverage and health risk factors in the United States. Am J Prev Med 22:75–83.
[19] Zukauskas G. (2004) The Suicide Formula. Publishing Centre of Mykolas Romeris University, Vilnius, 208p.
[20] Crawford MD, Gardner M.J, Morris JN. 1968. Mortality and hardness of local water-supplies. Lancet I: 827–831.
[21] Punsar S, Karvonen MJ. 1979. Drinking water quality and sudden death: observations from West and East Finland. Cardiology 64:24–34.
[22] Masironi R, Pisa Z, Clayton D. 1980. Myocardial infarction and water hardness in European towns. Environ Pathol Toxicol 4:77–87.
[23] Masironi R. 1987. Geochemistry, soils and cardiovascular diseases. Experientia 43:68–74.
[24] Brown JS Jr. 1994. Role of selenium and other trace elements in the geography of schizophrenia. Schizophr Bull 20: 87-398.
[25] Edelman E. 2001. Natural healing for schizophrenia and other common mental disorders. Third edition. Borage Books. 208p.
[26] Ranta J. Penttinen A. 2000. Probabilistic small area risk assessment using GIS-based data: a case study on Finnish childhood diabetes. Stat Med 19:2345–2359.
[27] Sliaupa S, Zukauskas G, Korabliova L, Denas Z, Zakarevicius A, Davidoniene O. 2004. Somatic diseases and human behavior in relation to geological factors. Juristprudencija 62:4-16.

Scientific Networking and the Global Health Network Supercourse
F. Linkov and R. LaPorte (Eds.)
IOS Press, 2006

© 2006 IOS Press. All rights reserved.

Road Traffic Accidents in Tunisia: A Man-Made Disaster

Hassen GHANNEM, MD, MSc
Department of Epidemiology, University Hospital Farhat Hached, 4000 Sousse, Tunisia (e-mail: Hassen.Ghannem@rns.tn)

Abstract. The objective of this paper is to determine the epidemiological profile and risk factors of traffic accidents in Tunisia in order to set up a national management strategy. We undertook a descriptive study, including all road accidents occurring throughout the entire territory of Tunisia during three consecutive years from 1996–1998. We studied a representative sample of 1,000 cases from the global database. The majority of accidents took place in large cities. Accidents in urban areas were four times more frequent than in the country; however, accidents occurring in rural areas led to eight time more deaths than in cities. Persons involved in road crashes were young, with an average age of 33.6 years. About 62% of the drivers were between 20 and 40 years old. Pedestrians were the most vulnerable type of road users, especially children and the elderly. Drivers' mistakes were inattention in 41.4% of accidents, followed by speed excess, and the disrespect of priority in 14.1% and 9.7% of cases, respectively. Inattention while crossing the street was the most frequent mistake of pedestrians (71%). Drivers used private vehicles in 50% of cases and motorcycles in second instances. The human factor was involved in 99% of traffic accidents, the environment in 9.3%, and vehicles in 4% of road accidents. We concluded that a national program of traffic accident prevention based on the education of young drivers is necessary to remedy this increasing public heath concern.

Keywords. Injury epidemiology, traffic accidents, transportation, Tunisia

Introduction

Road traffic accidents are rated 11th in mortality cases in the world, and constitute a major problem of public health [1]. They are the cause of death for 1.2 million people in the world, and are the cause of 50 million injuries every year [1]. The number of road accident victims is expected to increase by 65% between 2000 and 2020. This increasing rate is specific to developing countries [2]. Traumatisms represent 12% of the global burden of disease in the world. These traumatisms are dominated by road traffic accidents [3]. Road traffic accidents cost between 1 and 2% of the gross national product in every country [4], and represent a man-made disaster. Such accidents are greatly avoidable by the setting up road accidents prevention strategies. It is imperative that these strategies be based on a scientific approach to the problem. It is also important for these activities to come from a rational analysis based on the situation in every region [1]. Local data are essential for every country to adopt these prevention strategies [5]. The objective of this paper is to describe the epidemiological profile of road traffic accidents in Tunisia and discuss the potential for setting up preventive measures.

1. Material and Methods

1.1 Study Design

This is a descriptive study of the epidemiological profile of the road traffic accidents in Tunisia between January 1, 1996, and December 31, 1998.

1.2 Study Population

The study was conducted on a representative sample of 1,000 accidents (the sample size has been calculated on the basis of a prevalence of non-held seat belt estimated at 50% in serious accidents, with a precision of 3%, and a confidence level of 95%).

1.3 Data collection

These data concerned the following items:

- circumstances of the accident: geographical localization, chronology, infrastructure
- characteristics of transport means
- characteristics of the driver: age, sex, type of committed mistake, physical state
- characteristics of pedestrians involved

1.4 Source of Information

The source of information was the National Observatory of the Circulation that centralizes data concerning road traffic accidents taking place on the Tunisian territory.

2. Results

2.1 General Data

The general data for the background of the results is as follows:

- During 1996–1998, there were 32,197 accidents that generated 3,934 fatalities and 43,600 injuries, with 14. 9 deaths for every 100 accidents.
- The incidence of road traffic accidents was 116.3/100,000 inhabitants per year.
- The mean mortality rate was 14.16/100,000 inhabitants per year.
- The category of pedestrians was the most vulnerable and the most seriously affected: pedestrians represented 15.7% of the total number of people injured in accidents; 27% of injury cases; and 41.6% of deaths in relation to the road traffic accidents.
- People involved in the road traffic accidents were mostly young, with a mean age of 33.6 years. More than 61.7% drivers were between 20 and 40 years old.
- The majority of drivers involved in the accidents were male (21 times more than females).

2.2 Circumstances of the Accident

Localization

The majority of accidents took place in the large cities in the country. Accidents in towns were four times more frequent than in the countryside. However, accidents occurring in countryside led to eight times more deaths than in the cities.

Chronology

During the period of the study, the frequency of accidents was higher in August on Saturdays, with 10.4%; 15.2% of accidents occurred between 4 p.m.–6 p.m. Accidents that took place at the night between midnight and 6 a.m. were very dangerous with 8.6% of the total deaths.

2.3 Factors Involved in Accidents

Human Factors

Drivers' mistakes were inattention in 41.4% of accidents, followed by speed excess and the disrespect of priority 14.1% and 9.7% of cases, respectively. Not paying attention while crossing the street was the most frequent mistake of pedestrians (71%).

Factors Related to the Vehicle

Technical defects of the vehicle were found in 5.8% of accidents. There was a defect of light in 1.9% cases, and a bursting of wheel in 1.4% of cases.

Factors Related to the Road

In 50% of cases, accidents took place on local roads.

3. Discussion

Our study shows that human factors were involved in the majority of traffic accidents, the environment in 9.3% of cases, and vehicles in 4% of road accidents. Pedestrians were involved in terms of injuries and death related to the road traffic accidents. Our study has some limitations in relation to its retrospective design and the lack of important information about the vehicle and the degree of utilization of seat belts by passengers.

The mean mortality rate by road traffic accidents was 14.16/100,000 inhabitants/year in Tunisia between 1996 and 1998. Whereas, it was between 11–12/100,000 inhabitants/year in Western Europe; between 19.1 and 28.3/100,000 inhabitants per year in African countries [1]; these differences have been explained by the application of prevention strategies in the developed countries [1].

Pedestrians constitute the category most affected by the road traffic accidents in Tunisia. It is also the cause in developing countries [6]; whereas in the developed countries, car drivers constitute the category the most impacted by road traffic accidents [1]. Drivers involved in road traffic accidents in Tunisia are young men. This observation agrees with other national data [1].

There was a technical defect of the vehicle in 5.8% of the recorded accidents; in the world the contribution of the technical defects in vehicles in road traffic accidents was about 3% in the developed countries [1], and 5% in developing countries [7].

Although there were some limitations in our study, it still allowed us to collect some reliable data concerning road traffic accidents in Tunisia, which constitutes according to the international recommendations [1], a fundamental phase in developing a national program of traffic accidents' prevention based on education of young drivers to remedy this increasing public heath concern.

4. Conclusion

Road traffic accidents constitute a huge public health problem: a man-made disaster. The situation is going to be worse unless prevention strategies are adopted. There is hope to avoid such losses of lives, of health, and of money by setting road safety actions. A scientific approach to the issue is essential in every country so that road safety policies could be founded on reliable data, and meet financial resources of each region.

References

[1] Rapport mondial sur la prévention des traumatismes des accidents de la circulation. 2004. Genève (Suisse), Organisation mondiale de la santé.

[2] Nantulya VM, Reich MR. 2002. The neglected epidemic: road traffic injuries in developing countries. Br Med J 324: 1139-41.

[3] Rapport sur la santé dans le monde de 2001. 2001. La santé mentale: Nouvelle conception, nouveaux espoirs. Genève (Suisse), Organisation mondiale de santé.

[4] Jacobs G, Aeron Thomas A, Astrop A. 2000. Estimating global road fatalities. Crowthorne, Transport Research Laboratory (Rapport du TRL n° 445)

[5] O'Neil, Mohan D. 2002. Reducing motor vehicle crash deaths and injuries in newly motorizing countries. Br Med J 324: 1142-45.

[6] Brainard B. 1986. Injury profiles in pedestrian motor-vehicle trauma. Ann Emerg Med 18 : 881-3.

[7] Odero W, Garner P, Zwi AB. 1997. Road traffic injuries in developing countries: a comprehensive review of epidemiological studies. Trop Med Int Health 2: 445-460.

Scientific Networking and the Global Health Network Supercourse
F. Linkov and R. LaPorte (Eds.)
IOS Press, 2006

© 2006 IOS Press. All rights reserved.

Indication of Zinc Deficiency for Lithuanian Population and Its Geomedical Control

M. JAKUBENIENE[a], G. ZUKAUSKAS[a], S. SLIAUPA[b], J.SATKUNAS[c], I. KERIMOV[d]
[a]Institute of Forensic medicine of Mykolas Romeris University, S.Zukausko 12, LT-08234 Vilnius, Lithuania, [b]Institute of Geology and Geography, ,T.Sevcenkos 13, LT-03223 Vilnius, Lithuania, [c]Lithuanian Geological Survey, Konarskio 35, LT-03123 Vilnius, Lithuania, [d]Azerbaijan Academy of Sciences, Buil-Buil Avenue 14/70, AZ-1000 Baku, Azerbaijan

Abstract. Zinc is an essential bio-element that plays a fundamental role in a wide range of biochemical processes; this metal functions as a co-factor of multiple enzymes, it is a major component of various proteins and the modulator of the mammalian immune and nervous system. With zinc playing such a crucial role in normal biological and physiological functions, a deficiency of this mineral would be expected to result in a number of adverse physiological consequences. Symptoms resulting from zinc deficiency include infections, hypogonadism, weight loss, emotional disturbance, dermatitis, alopecia, delayed wound healing, etc. Many symptoms of zinc deficiency result from poor diet consumption, lack of zinc in food crops grown in zinc depleted soils, dependence on processed foods, but often the most severe symptoms result from other factors including excessive alcohol use, liver diseases, malabsorption syndromes, renal disease, and sickle cell disease.

The problem of zinc deficiency has been known for more than 40 years and a solution is still outstanding. In our days when many people are moving around the world with various missions, it is necessary to study the influence of natural geological factors on the possible zinc deficiency of population. On the ground of pilot investigation of zinc abundance in blood serum of patients with liver diseases (hepatitis and liver cirrhosis) and biological media of people with alopecia it was demonstrated that there are some marks of zinc deficiency in Lithuanian people. The causes of this conditioned zinc deficiency are largely explained in terms of excessive alcohol consumption and living in zinc depleted environment. The zinc deficient status can be considered as the cause of emotional disorder (mostly depression) inducing the suicide.

Keywords. Bioelements, zinc deficiency, environmental health, suicide

Introduction

During the first half of the 20th century, researches discovered that zinc was essential for the normal growth and survival of plants and animals [1; 2; 3]. Clear evidence of human zinc deficiency began to emerge during the 1960s, when Dr. Ananda Prasad first reported cases of dwarfism and delayed sexual maturity among Middle Eastern adolescents [4].

Further investigation of zinc status in organisms showed vital importance of this element for normal physiology and function of the body. Zinc functions as a required co-factor for over 200 zinc-dependent enzymes, which exert important influences on every major metabolic pathway, including the synthesis and degradation of carbohydrates, lipids, proteins and nucleic acids [5; 6]. Zinc is also an essential component of the endogenous storage forms of pancreatic insulin [7]. Zinc is required for each step of the cell cycle and is essential for DNA synthesis; the catabolism of RNA appears to be zinc-dependent [8]. A growing body of evidence implicates a derangement of zinc homeostasis in mood disorders [9].

With zinc playing such a crucial role in such an array of normal biochemical and physiological functions, a deficiency of this element would be expected to result in a number of adverse physiological consequences. Zinc deficiency in humans might manifest as severe, moderate, or mild [10]. The manifestations of severe zinc deficiency include bullous pustular dermatitis, alopecia, diarrhea, emotional disorder, weight loss, intercurrent infections due to cell-mediated immune dysfunctions, hypogonadism in males, neurosensory disorders, and problems with healing of ulcers. This condition can be fatal. Clinical manifestations of moderate and mild levels of zinc deficiency include growth retardation and male hypogonadism in adolescence, rough skin, poor appetite, mental lethargy, delayed wound healing, cell-mediated immune dysfunctions, abnormal neurosensory changes, oligospermia in males, and decreased lean body mass.

The symptoms of zinc deficiency can result from various factors [11]. When zinc deficiency status was first defined for Middle Eastern adolescents such fact was believed to be the result of low zinc soil levels in this region coupled with the widespread practice of consuming largely unleavened whole-grain breads rich in phytic acid that binds zinc [12]. The future investigation showed that some cultures (food crops) grown in the regions with depleted content of zinc in soil have decreased level of zinc and can result poor zinc diet consumption [13]. The most severe symptoms of zinc deficiency in humans can result from such factors as excessive alcohol use, liver diseases, malabsorption syndromes, renal disease, enteral or parenteral alimentation, and sickle cell disease [11].

We would like point out that clinical symptoms of zinc deficiency are quite various. Their manifestations differ for each person and as each organism is unique may result from different factors.

The level of general morbidity and mortality in Lithuania has increased significantly during last 10 years. There is the biggest suicide rate in Lithuania among European countries—44.5/100 000 inhabitants, nearly epidemic status with tuberculosis -86/100,000, quite high levels of alcohol delirium treatments, and illicit drug users [14; 15; 16; 17; 18]. During the last decade there was a significantly increased level of alopecia cases among young people [19]. Most of these diseases and such social malfunction as suicides, in some way, can be related with zinc deficiency status in the human body. It is known that a considerable percentage of suicide victims were suffering from depression [20]. On the other hand, zinc is involved in the pathophysiology and therapy of depression [9]. Moreover, the severity of depression (assessed according to Hamilton Depression Rating Scale) was negatively correlated with serum level of zinc [21; 22]. The causes of alopecia also can originate from zinc deficiency status of organism [23].

The aim of our study was to gain background information and conduct a pilot investigation of the zinc amount in blood serum of patients with liver diseases (hepatitis and liver cirrhosis) and biological media of people with alopecia to demonstrate that there are some marks of zinc deficiency in Lithuanian people. The causes of these conditions could partially be explained by excessive alcohol consumption and living in zinc depleted area (nutritional factor). The possible zinc deficient status of some Lithuanian people is considered as a cause of emotional disorder (mostly depression) that may induce such malfunction as suicide.

1. Data and Methods

1.1 Biological Specimens

Whole blood, blood plasma, urine, and hair in 63 patients ranging in age from 2 to 35 years with various cases of alopecia and blood serum of 56 patients with liver diseases (hepatitis C and liver cirrhosis) were examined.

1.2 Chemical Standards

Suprapure grade reagents and high purity water were used. Working standard solutions were prepared from FLUKA stock standards containing 1,000 mg/l of zinc, copper, lead, manganese and nickel by diluting with 5% nitric acid solution.

1.3 Analytical Methods

Atomic absorption spectrometer model AAS3 from Zeiss Jena, Germany, was used for determining the concentration of zinc. The continuous source (Deuterium lamp) was used to make a "background" correction. Method of standards addition was applied to compensate for matrix effects of biological specimens.

Samples for metals analysis were digested in the presence of concentrated nitric acid and perhydrol. Digests were diluted with the 5% solution of nitric acid.

1.4 Statistical Data

The rates of suicides in 45 administrative regions of Lithuania were selected from official statistical data about causes of death in Lithuania [14; 15; 16; 17]. The tools of geographic information system (GIS) were used presenting the distribution of rates of suicides in conventional maps of Lithuania. The data related to Baltic and Nordic countries were selected from publication of Nordic Medico-Statistical Committee [18].

The zinc content in soil of Lithuania was selected from the Biochemical atlas of Lithuania [24].

2. Environmental Aspects of Zinc in Lithuania

Zinc is a trace element that occurs in the soil in different forms, its abundance is strongly controlled by a rock composition. The geochemical studies of Lithuania indicate that zinc is a lithogenic element, this group shows the highest concentrations in the morainic and glaciolacustrine sediments (dominated by clay and fine-grained clastics), whereas indicating minimum abundances in fluvioglacial and aeolian sandy sediments [24]. Zinc also concentrates in the biogenic sediments (e.g., soil). Besides abundance in the soil, the mobility is an important parameter for the intake of this element. Zinc is mobile in the acid low-pH environment, composing soluble complexes in water, whereas the high carbonate content (e.g., north Lithuania) decreases its mobility. In Lithuanian soil, Zn has most close direct correlation with Cr, Cu, Ga, Sn, Ti, V, also La, Li, Nb, Rb, Sr.

Lithuania is characterized by the generally lowest concentration in the soil of the circum-Baltic region [29], which implies that currently the average Lithuanian diet does not contain sufficient amounts of this element because of depletion of vital minerals in the soil. The average content of zinc in Lithuanian soil is 20-30 ppm, while reaching 50 to 70 ppm in Sweden and Norway. Furthermore, Zn shows significant concentration variations across Lithuania from 5 to 150 ppm. On average, the lowest values are identified in southeast and east Lithuania (22 to 25 ppm in Lazdijai, Varena, Salcininkai, Svencionys, Moletai regions) that is mainly covered by aeolian and fluvioglacial sandy deposits (Figure 1). Also, low concentrations are identified in central and north Lithuania (25 to 28 ppm in Joniskis, Siauliai, Radviliskis, Kupiskis, Panevezys, Akmene regions). The highest average abundances are documented in the soil of west Lithuania (32 to 40 ppm in Plunge, Silale, Telsiai, Silute, Kretinga regions) and some other regions dominated by morainic and glaciolacustrine litholgies. These relationships are well reflected by statistical correlations; the correlation coefficient between zinc and fliovioglacial sediment percentage is -0.50, whereas +0.40 with glaciolacustrine sediment percentage.

The surface lithologies are intimately related to geomorphic features and other landscape parameters that result in respective relationships with zinc. Specifically, the forested areas (percentage) have very close correlation with zinc, the coefficient is as high as -0.43. This relationship is of double nature (i) the sandy soil is more extensively used for forestry, and (ii) the more acid environment, typical for the forest areas, increases mobility of zinc.

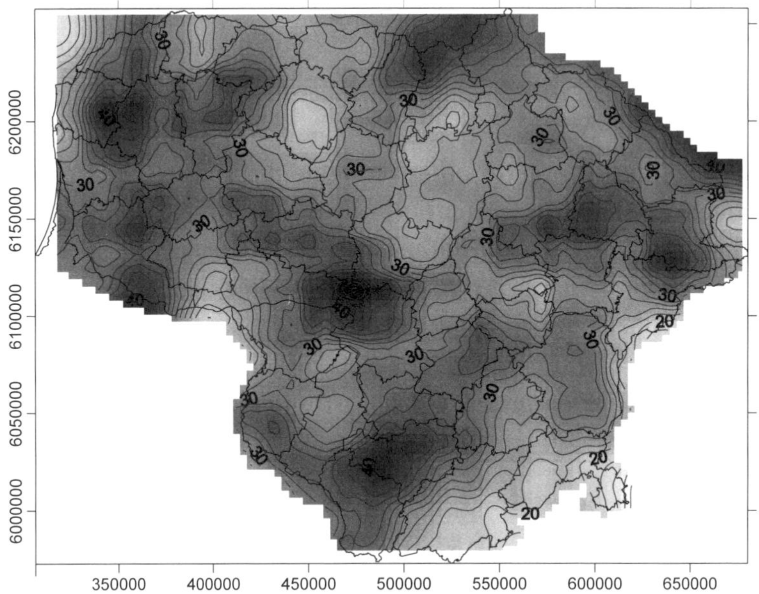

Figure 1. Zn concentrations (ppm) in the soil of Lithuania (low-pass filtered).

3. Results

Zinc deficiency in humans is related to nutritional factors (consumption of some vegetable food grown in region with depleted content of zinc in soil), excessive intake of alcohol, cirrhosis of the liver, malabsorption syndrome, chronic renal disease, burns, iatrogenic causes, diabetes, and genetic disorders [10]. The evaluation of all these factors might be an objective of extensive study establishing the zinc status of individual person and further on the zinc status of concrete population. In the framework of this report, we will try to evaluate only some of them: the nutritional factors (not directly, but on the ground of such geochemical factor as zinc content in soil); the influence of excessive intake of alcohol on the development of liver diseases and the zinc status of patients with liver cirrhosis. Finally, we will try to evaluate the zinc status of people suffering from such disease as alopecia.

3.1 Nutritional Factor

The major source of the body's trace elements is the soil. There are various pathways through which the trace elements find their way into the body. The concentrations of trace elements in rocks vary by rock type and location and are a fundamental control on the availability of trace elements to humans. Trace-element concentrations are modified by a variety of natural processes and deliberate and accidental human activities. Agricultural chemicals and pollutants may be added. Crops selectively remove from the soil the elements they require for growth. The drinking water contains trace elements leached from rock and soil and may also have been polluted or chemically treated.

The relation of content of zinc in soil as the possible cause of zinc deficient diet that can result in a zinc deficiency state in an organism and rates of suicides in 45 different regions of Lithuania, as well as in Baltic and Nordic countries, was investigated. Lithuania is country that has one of the world's highest suicide rates. Around 1,500 people commit suicide every year [20]. Figure 2 presents the average rates of suicides in different regions of Lithuania for four years (1999, 2001, 2002, 2003).

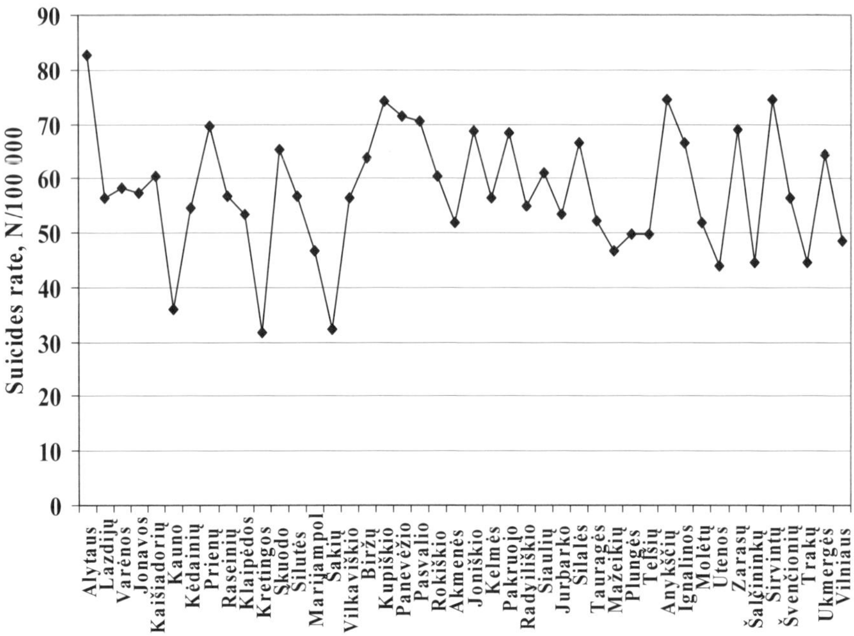

Figure 2. The rates (events of death per 100.000 habitants) of suicides in different regions of Lithuania (average rate for four years:1999,2001,2002,2003).

This distribution shows that the rates of suicides differ considerably among various regions of Lithuania. When we calculated the average rate of suicides for 45 different regions, the average rate of 57.6/100,000 was obtained. The average rate of suicides in 11 regions of Lithuania (Lazdijai, Varena, Salcininkai, Svencionys, Moletai, Joniskis, Siauliai, Radviliskis, Kupiskis, Panevezys, Akmene) with depleted content of zinc in soil was discovered to be 63.2/100,000. Alternatively, the average rate of suicides in regions with the highest zinc abundance in soil (Plunge, Silale, Telsiai, Silute, Kretinga) was obtained at 51.2/100,000. Comparing these data we can conclude that the rate of suicides in regions with depleted content of zinc in soil exceeds the average rate of suicides per Lithuania, and considerably exceeds rate of suicides in regions with highest zinc abundance in soil.

3.2 Excessive Alcohol Intake and Liver Diseases

Excessive alcohol consumption is mentioned as one of the factors resulting in zinc deficiency in the organism. Although alcohol affects a number of body organs, it is essentially harmful to the liver. Alcohol is metabolized in the body, and the liver performs most of the work, potentially incurring serious damage in the process. Not only does alcohol destroy liver cells, it also destroys their ability to regenerate, leading to a syndrome of progressive inflammatory injury to the liver. Alcoholic inflammation of the liver will often eventually progress to cirrhosis.

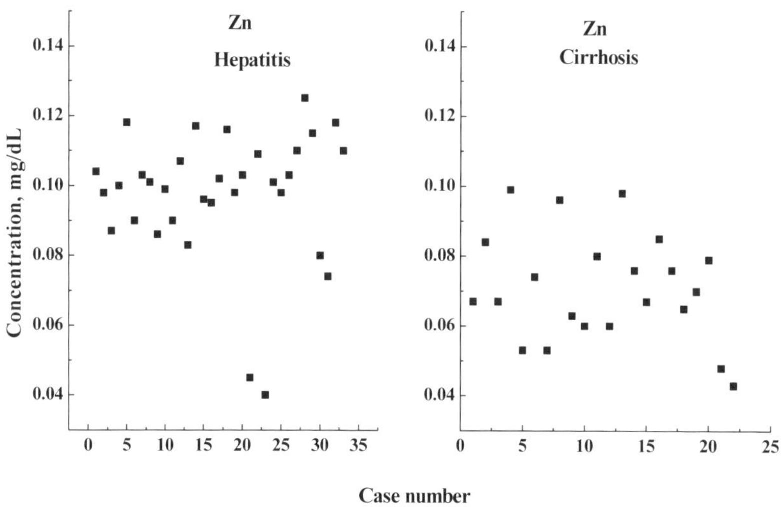

Figure 3. Zinc serum concentration in patients with hepatitis (left picture) and liver cirrhosis (right picture).

The first documentation of the abnormal zinc metabolism that occurs in patients with alcoholic cirrhosis was provided by Vallee (Vallee 1956). Today, there is some variance of opinions as to how often such phenomena occur in cirrhosis and other liver disease.

Serum zinc concentrations were investigated in n=33 patients with hepatitis (C) and in n=22 patients with liver cirrhosis. Figure 3 presents the results of this study.

3.3 Zinc in People with Alopecia

Approximately 1% of the population will have had alopecia areata by the age of 50 and the peak incidence occurs in children and young adults [25]. Alopecia areata has been associated with atopic dermatitis, autoimmune disease, vitiligo, and endocrine disease. The intoxication by metal compounds in many medical references is mentioned as one of possible cause of alopecia [26; 27]. On the other hand, it is known that alopecia may be related to the deficiency of essential metals such as zinc, copper and iron [26; 28].

The concentration of zinc was investigated in 63 cases of alopecia. The occurrence of various cases of alopecia according to the sex of patients in different aged group is presented in Table 1.

Table 1. Occurrence of various cases of alopecia in patients of different age groups

Age of patients	**Sex of patients**	
	Male	**Female**
2-5 years	9 cases: 1- Alopecia Totalis 8- Alopecia Areata	4 cases: 4-Alopecia Areata
7-13 years	10 cases: 1- Alopecia Diffusa 9- Alopecia Areata	15 cases: 2 - Alopecia of eyebrows and eyelashes 13- Alopecia Areata
14-20 years	6 cases: 2 - Alopecia Totalis 1 -Alopecia Diffusa 1 -Alopecia of eyebrows and eyelashes 2 - Alopecia Areata	1 case: 1- Alopecia Areata
21-35 years	6 cases: 1 - Alopecia Totalis 1 -Alopecia Diffusa 3- Alopecia Areata	12 cases: 1-Alopecia Diffusa 11- Alopecia Areata

The distribution of alopecia cases according to the age of patients shows that the largest number of cases began in the period that is called "the secondary infancy," i.e., between the ages of 6 and 12 years, and corresponds with data of others authors [28].

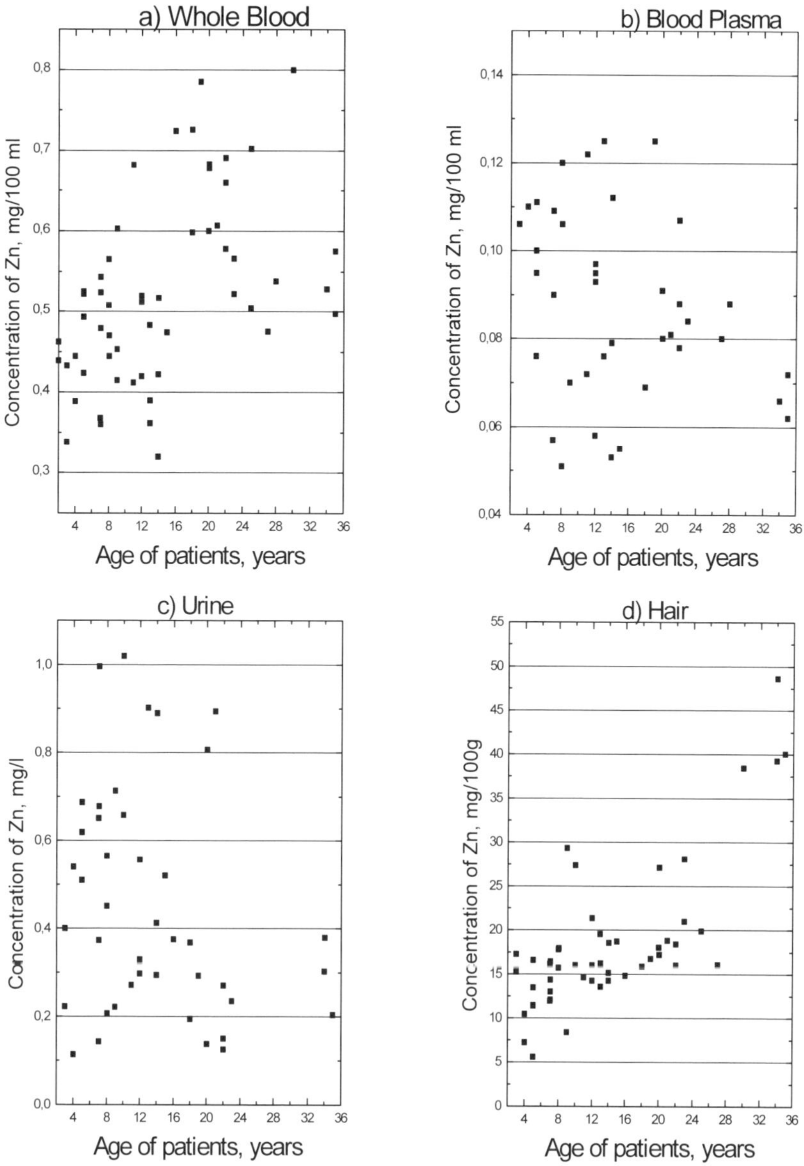

Figure 4. Concentration of zinc in various biological media accordingly to the age of patients

In our practice, patients with alopecia areata were investigated in most cases. It is related to psychosocial problems, which arose for young people with alopecia areata. The etiology and pathogenesis of alopecia areata is still entirely unknown. This disease runs a chronic recurrent course usually without any symptoms, whatsoever, and it tends to spontaneous recovery [26]. The data from Table 1 show that alopecia areata occurs even in infancy. We investigated the alopecia areata in 2-year old children.

The results of analysis of zinc in various biological specimens of patients with alopecia areata depending of the age of patients are shown in Figure 4.

The evaluation of the results for zinc in biological media is difficult. First of all, he dependence of concentrations of zinc in various biomedia according to the age of patients is unknown. Berfenstam (1952) specifies that in newborn infants the zinc content of the erythrocytes is only one-quarter of the adult values, rising progressively over the first 12 years of life (Prasad et al. 1975) indicated that plasma zinc levels in the newborn are in the same range as in adults. The levels fall to just below the adult level within the first week of life and continue to decline until three months, finally reaching the adult level at four months of age. So we can hypothesize that concentration of zinc in the whole blood could rise progressively at least over the first 12 years of life. However, the exact determination of zinc level in blood according to the age of patients could be the objective of future research studies.

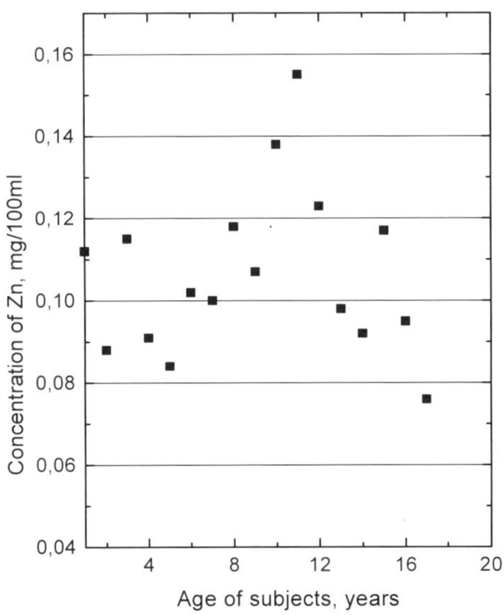

Figure 5. The concentration of zinc in blood plasma of persons from control group

The results of zinc investigation in blood plasma of subjects from the control group (persons who do not have problems with hair and feel healthy) are presented in Figure 5.

The comparison of zinc concentration in plasma for patients with alopecia areata and persons from the control group showed that the concentration of zinc in plasma of patients with alopecia areata in some cases is markedly lower than in persons from the control group. The plasma zinc values have a tendency to decrease with the age of patients and can be related with some conditional deficiency of zinc.

The examination of blood zinc in patients with alopecia areata show that blood zinc levels in the most investigated cases were lower than range presented by medical sources 0,650 – 0,880 mg/100ml (29; 30; 31].

4. Discussion

According to the report of International Zinc Nutrition consultative Group (IZiNCG; www.izincg.ucdavis.edu), one in five people worldwide lack enough zinc in their diet. Long considered impossible, it has been calculated that a substantial section of the population of the United States are at risk of suboptimal zinc nutrition [32]. The investigation of Zn deficiency on the behavioral effects in Chinese and Mexican-American children showed beneficial effects of Zn repletion on its neuropsychological function. It was given that research on cognition, behavioral activity, and brain electrophysiology as outcomes of Zn deficiency, and response to improved Zn nutrition was critical. So it could be concluded that Zn deficiency is common in both developing and developed countries [33]. Based on the results of our study, we can hypothesize that there are some zinc deficiency marks in Lithuanian population.

First of all, the most worrisome question for the Lithuanian people is why the rate of suicides in our country is the highest in the world. Making an attempt to answer this question we would like to point out several interdependent risk factors. As it was mentioned, zinc deficiency status of an organism is strongly dependent on nutritional factors and excessive intake of alcohol. The alcohol consumption in our country is greatest among Baltic and Nordic countries [18]. As a consequence, the rate of liver diseases in Lithuania exceeds the rates neighboring countries (Figure 6).

Our study measuring zinc concentration in patients with liver cirrhosis (the most common cause of liver cirrhosis is alcohol abuse) and patients with hepatitis showed that patients with liver cirrhosis had a depressed content of zinc in their blood serum. This result confirmed the data of others researches that alcoholic liver cirrhosis is associated with zinc deficiency [34; 35].

The monitoring study of some regions of Lithuania [36] displays that zinc concentration in children and pregnant women does not exceed 0.480 mg/dL in a Siauliai resident and 0.380 mg/dL in a Vilnius resident. It is difficult to evaluate these data because of lack of the normal permissible values of zinc in biomedia for the Lithuanian population, but comparing this data with normal concentration from references [37; 38; 39] it seems to be deficient.

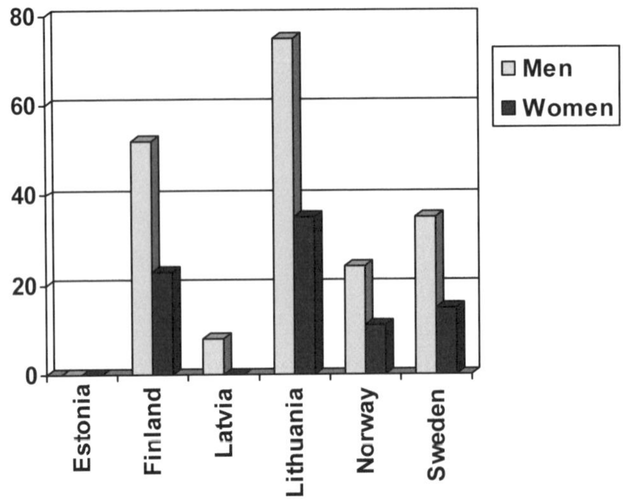

Figure 6. Rates of alcohol liver disease in 2002; Data for Estonia could not be divided by sex; the total rate for men and women was 37.8.

The investigation of zinc in biomedia of people with alopecia showed that the zinc in blood plasma of these patients is decreased compared to the control group. This result coincides with the data obtained in another investigation of zinc in people with alopecia [40].

Numerous studies have shown that a variety of mental and behavioral changes have also been associated with zinc deficiency in humans including apathy, lethargy, amnesia, irritability, depression, paranoia, as well as mental retardation [36; 41].

All of these emotional disorders are directly related with suicides [20]. So we can find some relation between the possible zinc deficiency status of the Lithuanian population and high rates of suicides in Lithuania. The possible nutritional factor also can be confirmed comparing the rate of suicides in various regions of Lithuania with a different content of zinc in soil. It is only our hypothesis, because the establishment of zinc status requires more detailed monitoring study of the whole population. Additionally, what must be evaluated is the interaction of zinc in the organism with other elements and the influence on health status including such geo-factors as gravity and magnetic field, sub-soil lithologies, relief, some other landscape feature, drinking water, and soil chemistry. It is also very important to compare obtained results with data for populations of neighboring countries. The answers to these questions will be the objective of future studies.

5. Conclusion

Though a trace metal like zinc is, in weight terms, only a minuscule part of the human metabolism, its presence is absolutely vital in all the major metabolic pathways. On the basis of current evidence it is necessary to initiate a more detailed investigation of the zinc status in the human body. It will be done by evaluating other factors influencing this status: interaction of zinc in the organism with other elements, and influence on health status including such geo-factors as gravity and magnetic field, sub-soil lithologies, relief, some other landscape feature, drinking water and soil chemistry. Due to the fact that 20 percent of the world's population suffer from a zinc deficient diet, it is essential to know how to improve this condition. Knowing more about this condition may have crucial significance for human outcomes, behavioral health, and mental conditions.

The results of our zinc study investigation in biological media of patients with alopecia areata and liver cirrhosis can be useful as experimental evidence verifying zinc status in the Lithuanian population.

References

[1] Sommer AL 1928. Further evidence of the essential nature of zinc for the growth of higher green plants: The search for elements essential in only small amounts for plant growth. Science 66 482-484.

[2] Todd WR, Elvehiem CA, Hart EB. 1934. Zinc in the nutrition of the rat. Amer. J. Physiol 107:146-156.

[3] Tucker HF, Salmon WD. 1955. Paraketarosis or zinc deficiency disease in pigs Proc.Soc.Exp.Biol.Med 88:613-616.

[4] Prasad AS, Rabbani P, Abbash A. 1978. Experimental zinc deficiency in human, Ann Inter Med 89: 483.

[5] Agget PJ. 1985. Physiology and metabolism of essential trace elements: an outline, Clinics Endocrinol Metab 14: 513-543.

[6] Underwood E. 1977. Trace elements in human and animal nutrition, 4th edition, New York: Academic Press.

[7] Murray M. 1998. Encyclopedia of Natural Medicine,Second ed. Rocklin, CA: Prima Publishing.

[8] Prasad AS. 1983. Zinc deficiency in human subjects. Prog Clin Biol Res 129:1-33.

[9] Nowak G, Szewczyk B. 2002. Mechanisms contributing to antidepressant zinc actions. Pol J Pharmaco 54: 587-592.

[10] Prasad AS. 1988. Zinc in growth and development and spectrum of human zinc deficiency. J Am Coll Nutr 7:377-384.

[11] Evans GW. 1986) Zinc and its deficiency diseases. Clin Physiol Biochem 4: 94-98.

[12] Rossander L., Sandberg AS, Cederblad A. 1992. Absorption of zinc from lupin (lupinus angustifolis) based foods. Br J Nutr 72: 865-871.

[13] Kalaici M, et al. 1999. Grain yield, zinc deficiency and zinc concentration of wheat cultivars grown in zinc deficient calcerous soil in field and greenhouse Field Crops Research 63: 87-98.

[14] Causes of death. 2000. Statistics Lithuania. Statistica, Vilnius.

[15] Causes of death. 2002. Statistics Lithuania. Statistica, Vilnius.

[16] Causes of death. 2003. Statistics Lithuania. Statistica, Vilnius.

[17]. Causes of death. 2004. Statistics Lithuania. Statistica, Vilnius.

[18] Health statistics in the Nordic countries 2002. 2004. MONESCO.

[19] Jakubeniene M. 2000. Metals concentration in the blood, urine and hair in young people with alopecia. Vilnius, Fifth Baltic Congress of Laboratory medicine, Vilnius, May 4-6.

[20] Zukauskas G.(Ed.) 2004. The suicide formula: collective scientific study. Vilnius.

[21] Schlegel-Zawadska M, Zięba A, Dudek D, Krosniak M, Szymaczek M, Nowak G. 2000. Effect of depression and of antidepressant therapy on serum zinc levels – a preliminary study. Trace elements in man and animals 10, Kluwer Academic Plenum Pres 607-610.

[22] Skalnyj AV. 1999. Microelemental diseases (diagnostic and treatment). Scientific World, Moscow.
[23] Kadunas V, Budavicius G, Gregorauskiene V, Katinas V, Kliaugienen V, Radzevicius A, Taraskevicius R. 1999. Geochemical Atlas of Lithuania, Vilnius.
[24] Sahn EE. 1995. Alopecia areata in childhood. Semin Dermatol 14: 9-14.
[25] Gailevichius P. 1995. Plaukų lygos. Vilnius: Avicena.
[26] Kostanecki W. 1996. Choroby wlosow. Warszawa: Panstw. Zakl. Vyd. Lekarsk..
[27] Hair grows and hair regeneration. 1959. Annals of the New York Academy of Sciences. 83: 359-512.
[28] Prasad AS, et. al. 1975. Role of zinc in man and its deficiency in sickle cell disease. Erythrocyte structure and Function, Progress in Clinical and Biological Research; 1:603-619.
[29] Rose G.A, Willden EG. 1972. Whole blood, red cell, and plasma total and ultrafilterable zinc levels in normal subjects and in patients with chronic renal failure with and without hemodialysis. Brit J Urol 44: 281-286.
[30] Iyegar GV, et al. 1978. The elemental composition of human tissues and body fluids. NY:VERLAG CHEMIE.
[31] Sandstead HH. 1973. Zinc nutrition in the United States. Amer J Clin Nutr 26:1251-1280.
[32] Penland JG. 2000. Behavior data and methodology issue in studies of zinc nutrition in humans. J Nutr 130: 361S-364S.
[33] Taniguchi S, Kaneto K, Hamada T. 1995. Acquired zinc deficiency associated with alcoholic liver cirrhosis. Int J Dermatol 34:651-652.
[34] Scholmerich J, Lohle E, Kottgen E, Gerok W. 1983. Zinc and vitamin A deficiency in liver cirrhosis. Hepatogastroenterology 30: 119-125.
[35] Ptashekas J, Ciuniene E, Barkiene M, Zurlyte I. et al. 1996. Environmental and health monitoring in Lithuanian cities: exposure to heavy metals and benz(a)pyrene in Vilnius and Šiauliai residents. J Envir Pathol Toxicol Oncol 15:135-141.
[36] Naginiene R, Abdrachmanovas O, Kregzdyte R, Ryselis S. 2002. Investigation of heavy metals in people with alopecia. Trace Elements and Electrolytes 19: 87-90.
[37] Sprinivasan D, Marr S, Wareign R, et al. 1982. Magnesium, zinc and copper in acute psychiatric patients, Magnesium Bulletin 1:45-48.
[38] Berfenstam R. 1952. Studies of blood zinc. Acta Paediat 41: 389-391.
[39] Hair and hair diseases. 1990. Berlin: Springer.
[40] Maes M, D'Haese PC, Scharpe S, D'Hondt PD, Cosyns P, De Broe ME. 1994. Hypozincemia in depression J Affect Disorders 31: 135-140.
[41] Nowak G, Kubera M, Maes M. 2000. Neuroimmunological aspects of the alterations in zinc homeostasis in the pathophysiology and treatment of depression. Acta Neuropsychiatr 12: 49-53.
[42] Prasad AS. (ed.) 1966. Metabolism of zinc and its deficiency in human subjects. Springfield, Illinois:Zinc Metabolism Charles C Thomas; 250-302.
[43] Reimann C, Siewers U, Tarvainen T. et al. 2003. Agricultural soils in northern Europe: a geochemical atlas. Geochemisches Jarbuch Sonderhefte. Reihe D, Heft SD5.
[44] Toxicological profile for zinc. 1994. U.S.Department of Health & Human Services: Public Health Service, Agency for Toxic Substances and Disease Registry. TP-93/15.
[45] Vallee BL, Wacker WEC, Bartholomay AFF, Hoch L. 1956. Zinc metabolism in hepatic disfunction. I. Serum zinc concenytation in Laënnec's cirrhosis and their validationequential analysis, N (45). Engl J Med 255: 403-408.

Scientific Networking and the Global Health Network Supercourse
F. Linkov and R. LaPorte (Eds.)
IOS Press, 2006

© 2006 IOS Press. All rights reserved.

Environmental Risk Assessment

L.D. GALATCHI
Ovidius University, Department of Ecology and Environmental Protection, Mamaia Boulevard 124, RO-900527 Constanta - 3, Romania; e-mail: galatchi@univ-ovidius.ro and liviugalatchi@yahoo.com

Abstract. Environmental risk assessment evaluates the quantitative and qualitative characteristics of the environment in order to highlight the risk on environment and human health due to the potential presence or use of specific pollutants. Risk is a probability of an adverse direct or indirect effect on the environment or human health. It is a combination of the probability of occurrence of an event and the possible extent of that event's adverse effects and consequences, in terms of adverse effects on the ecosystem and human injury. Risk is defined as the probability of an event to occur related to the seriousness and extent of its consequences. An environmental risk assessment should be conducted, by adopting a systematic approach, when it is determined that a management action may have consequences to either the state of the environment or human health or well-being.

Keywords. Communications, environmental health, public health, risk assessment

Introduction

Ecological risk assessment and human health risk assessment are the two related disciplines from which environmental risk assessment is comprised. Both the ecological and the human health risk assessment, involves four steps:

- determination if a chemical, physical, or biological entity could induce adverse effects on individuals, populations, communities, or ecosystems being linked to a particular effect on the environment (*hazard identification*),
- determination of the relationship between the magnitude of the exposure and the probability of occurrence of the effects (*dose-response assessment*),
- determination of the extent of exposure, of the condition under which an organism comes into actual contact with a stressor (*exposure assessment*),
- description of the nature and the magnitude of risk (*risk characterization*).

1. The Interactive Nature of Environmental and Human Health Risk Assessment

There are several successive interactive stages at the core of the environmental risk assessment. Problem identification is the most important activity describing the resources that could be affected, and the possible consequences of the action.

The first tier is the scoping activity, which identifies concerns and issues that have potential risks. Further detailed studies are done if there are identified unacceptable risks.

Related to the levels of analysis, system boundaries (Figure 1) and risk expressions (Figure 2) through which the scoping activity proceeds; there are some questions to be asked.

The second tier is screening in which the data describing the dose-response relationships of exposure scenarios is analyzed. This is the stage where the risk is identified.

During the third tier, modelling of exposure and direct measures of effects are performed [1].

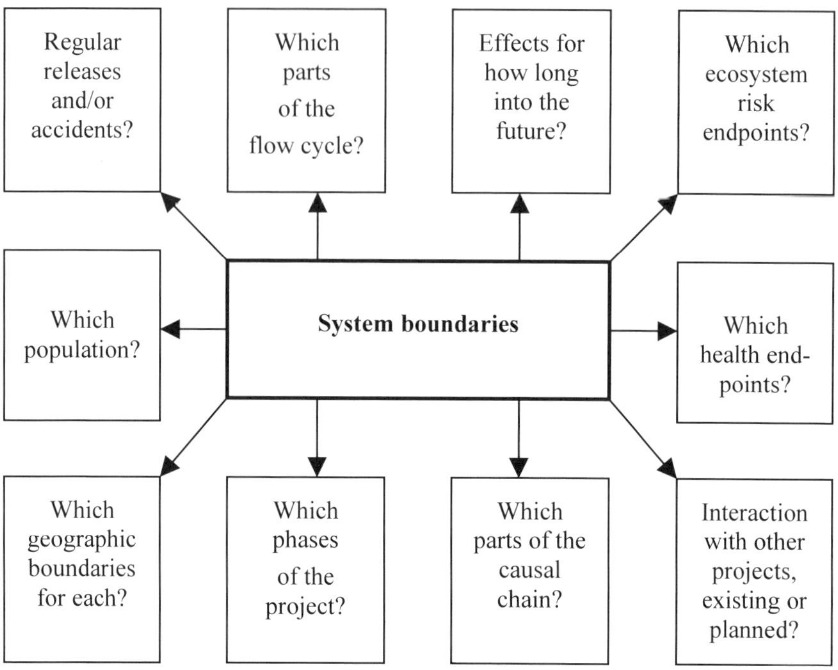

Figure 1. Questions related to system boundaries.

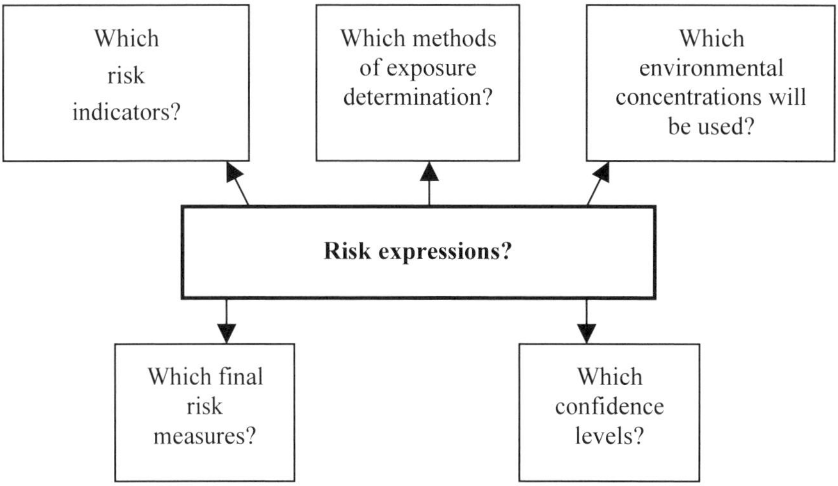

Figure 2. Questions related to risk expressions.

2. Risk Characterization and Comparison

Risk characterization aims to organize the information gained through the two main actions performed in environmental risk assessment: identifying and reducing the risks (Figure 3).

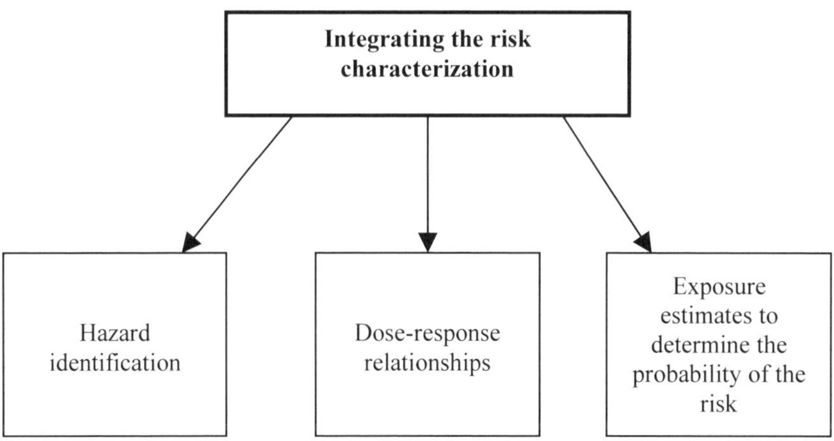

Figure 3. Information integrated by the risk characterization.

The risk characterization should also contain a description and estimate of the uncertainty of the assessment. It is recommended to include different points of views when performing the environmental risk assessment (Figure 4).

Useful comparison could be related to risks supposed by alternative technologies, sites, or projects that might provide equivalent economic growth benefits. Also, important to reference points to interpret new risks, if they are comparable, are the familiar risks. When the development projects are large enough to offer benefits, the risk may be acceptable. One of the issues to be considered from comparison of the risks is the cost/benefit of the risk reduction.

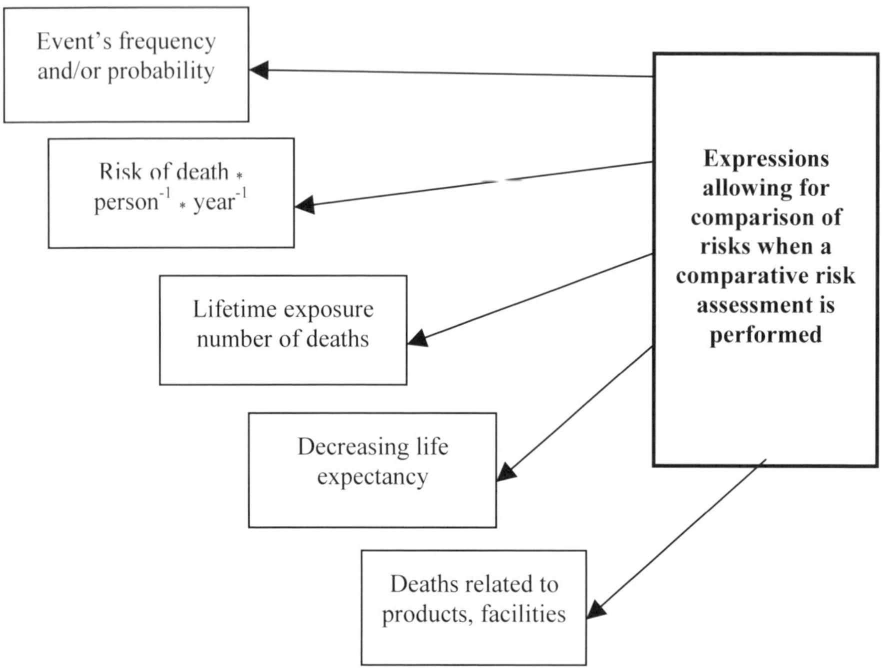

Figure 4. Expressions allowing for comparison of risks

3. Communication and Management in Risk Assessment

The reaction of the public to a risk that has been recognized, analyzed, and categorized depends upon how they perceive and understand the information (Figure 5). This is why decision analysis, through which the risk information is communicated, should be formed by analyzing available options, risks, costs, benefits, and their distribution within the society.

Risk management involves an evaluation of cost-effectiveness of risk reduction measures so that priorities can be made. Evaluating the measures in order to reduce the risk and implementing the cost-effective ones is needed to manage the risk.

There are three important phases of managing the risks (Figure 6). The phases show the necessity of identifying the hazard to people and environment, the quantification of the possibility for such hazards to occur, the magnitude of events, and then it is necessary to establish the limits of risk acceptability.

The last phase consists of the design and implementation of measures of risk control.

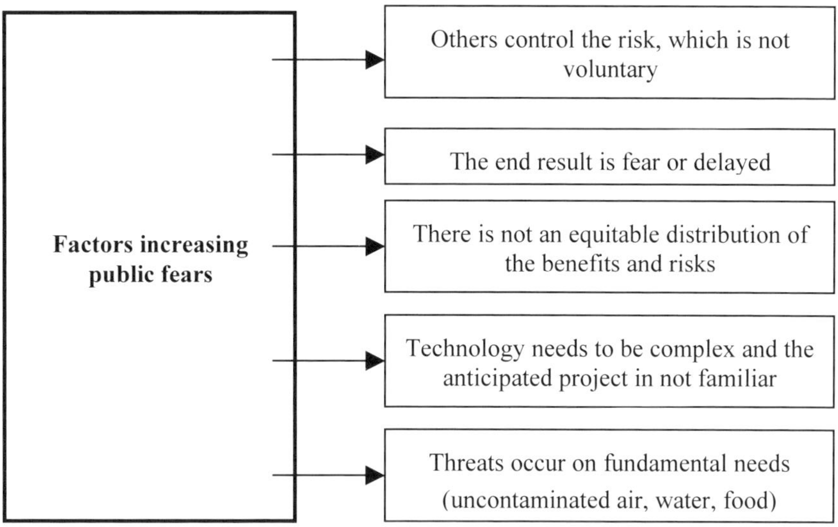

Figure 5. Factors that can lead to strong opposition to projects; this figure refer to false public perceptions and the need to engage in effective risk communication

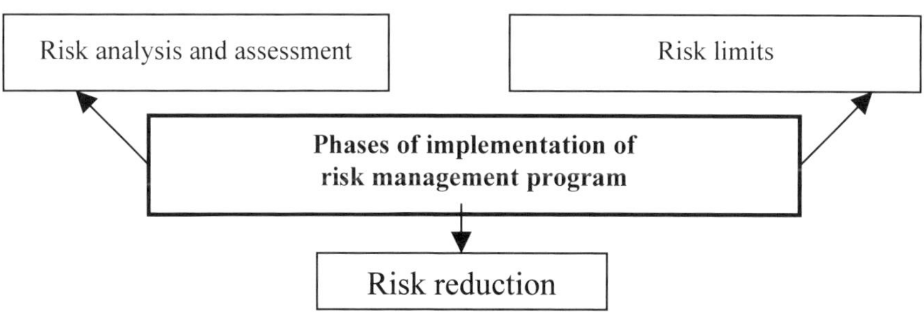

Figure 6. The three phases of the risk management.

4. Assessing the Environmental Risks

A risk assessment comprises three steps: hazard identification, consequence analysis, quantitative analysis and frequency assessment (Figure 7).

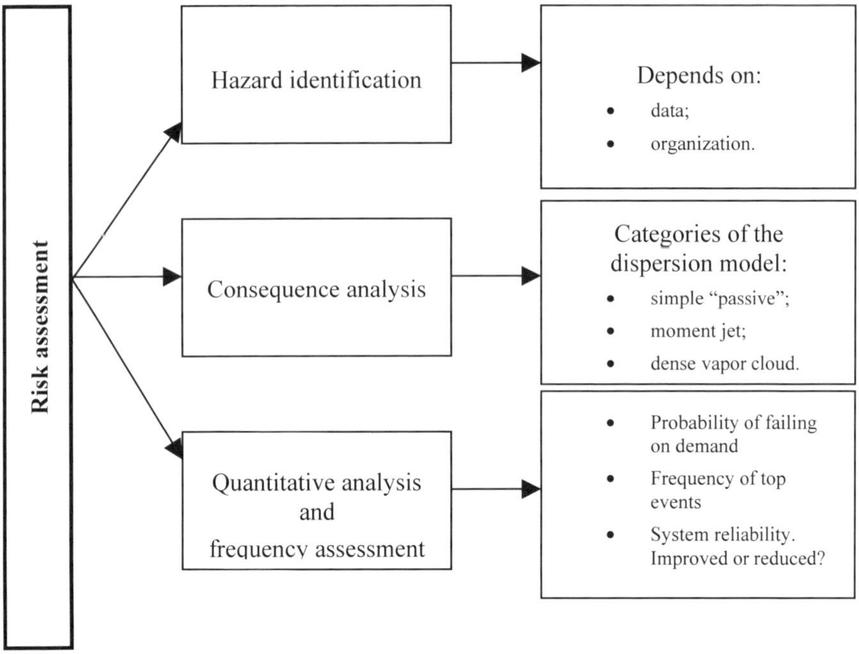

Figure 7. Steps of the risk assessment. This introduces a management issue concerning a chemical release to the environment (an industrial accident).

An emergency plan is supposed to comprise two components, on-site measures and off-site measures.

The on-site emergency plan includes:

- the required mechanisms for alarm and communications ;
- appointment of qualified personnel
- details of the emergency control centers

The off-site emergency plan should be based on these features: organization, communications, specialized emergency equipment and knowledge, voluntary organizations, chemical and meteorological information, humanitarian arrangements, public information, and assessment (Romaniei 2001) [2].

Acknowledgements

We thank Professor Stoica-Preda Godeanu for helpful suggestions during our research.

References

[1] Vadineranu A.2000. Dezvoltarea durabila. Bucharest: University of Bucharest Press; 1-224 p.
[2] Romaniei G. 2001. Bucharest: Strategia Nationala pentru Dezvoltare Durabila; 1-120 p.

F. Linkov and R. LaPorte (Eds.)
IOS Press, 2006
© 2006 IOS Press. All rights reserved.

Examination of Narcotic Drugs and Ethyl Alcohol among Soldiers in Lithuania

Robertas OBRIKIS[a], Darius JANKAUSKAS[b]
[a, b]*Siauliai Expert Department of the Lithuanian Institute of Forensic Medicine, Military Medical Service*

Abstract. Lithuanian young people are currently facing problems related to narcotic drugs and alcohol addictions. In order to fight against the spread of narcotic addiction in the country, it is necessary to identify people who are using narcotics, toxic drugs, and medicines. The purpose of this research study is to ascertain if soldiers have any opportunities to abuse the aforementioned substances while serving in the Lithuanian army, and to evaluate whether they know only the names of narcotic drugs or have more information about their use. Twenty-five cases of ethyl alcohol and narcotic drug use have been determined among 318 soldiers, who have been tested since 2002 (it composes 7.86% of the sample). There are no conditions for such harmful habits in the army.

Keywords. Alcohol addiction, drug addiction, military health

Introduction

Rapid changes in economic and social spheres are affecting the lives of the Lithuanian people. Among these changes are drug abuse—a very serious problem that goes side-by-side with other common poor habits, such as smoking and alcohol abuse. Today, obtaining narcotic drugs is very easy in Lithuania; the supply of illegal narcotic drugs is increasing with increased demand for such products.

Narcotic drugs could be described as natural or chemical substances that are included in the approved lists of controlled substances by the Ministry of Health. Use of such substances causes severe health disorders due to their harmful side-effects evoking psychological and physical dependence or danger to health [1; 2; 3].

Young people are vulnerable to such substances because they are not afraid to take risks and are curious; therefore, they often feel the need to prove themselves. Later on, young people may start using narcotic drugs or even become engaged in dangerous activities related to narcotic drug abuse. Research of high school students ages 15–16, has shown that the number of children who smoke, use alcohol, and narcotic drugs is increasing [4].

1. Statistics

Since 2001, 69,968 patients who suffer from dependence have been registered in the medical care institutions. 4,087 of them are addicted to narcotic drugs. Distribution of addicts according age is the following:

- 14 years old and younger—0.1% (5)
- 15–19 years old—11.3% (461)
- 20–24 years old—28% (1,142)
- 23–24 years old —37% (1,513)
- 35 years old and older—23.6% (966)

Distribution of addicts according gender and residence:

Towns – 94%	Male – 81.5%
Villages – 6%	Female – 18.5% [4]

The majority of soldiers in the Lithuanian army are young people who cannot avoid problems that are typical to our society. For example, there is the Department of Narcotic Drugs and Alcohol Control in United States Army, which is responsible for executing tests for the use of alcohol and illegal substances. Examination in U.S. Army is executed using general and selective methods. If the test results turn out to be positive, a soldier cannot serve in the army for six months. They are referred for further care [1].

2. Research Target

In order to decrease the use of narcotic drugs and alcohol in the country, it is necessary to identify who are abusing alcohol, narcotic and toxic drugs, and other medicinal substances.

Epidemiological examinations executed among the recruits of the Lithuanian army and soldiers of the initial compulsory army service determined that the problem of using narcotic drugs and alcohol exists in some departments of the army [1]. Using the aforementioned substances can potentially have a detrimental effect on the physical and mental health of soldiers. It can cause problems related to dependence or addiction, making the soldier unable to carry our his commitments and military duties.

When necessary, using psychological examinations, laboratory blood and urine analysis and psychiatric testing, the military medicinal committee can determine if the conscripts have a dependence on alcohol, narcotic drugs, or other substances that have an influence on the psyche. If alcohol dependence or dependence on narcotic drugs is determined, a conscript is judged to be unfit for army service.

The purpose of our research is to determine whether soldiers have any opportunities and conditions to abuse the aforementioned substances, and if they use them.

2.1. Research Object

From 2002 through 2004, neurological reactions as well as blood and urine toxicological analysis of 318 soldiers in Lithuania army have been evaluated. Soldiers belonging to the supervised group were tested first. These are persons who are either under major risk of using narcotic drugs or those who have some disorder because of using narcotic or psychoactive substances, but who were adjudged as proper to serve in the army [1]. Soldiers who have returned from short-term leave, public events, or entertainment places (discos, cinemas, parties) organized outside the limits of the unit, were randomly selected for testing. Soldiers, who have obvious drug dependence problems and also those with minimal usage but already noticeable symptoms, were tested by a symptomatic method, based on the suspicion of a medical officer of the subunit, and on noticing physical, emotional, or psychological changes in the behavior of a soldier during practical training and daily life.

Table 1. Partition of tested soldiers into groups

Groups of tested soldiers	Cases	Per cents (%)
Supervision group	20	6.29
Randomly selected group	272	85.53
Symptomatically selected group	26	8.18
TOTAL	**318**	**100**

Testing means identifiying whether a person uses narcotic drugs and alcohol. It is a reliable medical method to determine which kind of substances are being used.

According to the 10th revision of the International Statistical Classification of Diseases and Related Health Problems [5], the following groups of substances that affect psyche have been determined during testing:

- ethyl alcohol
- opiates
- cannabinoids (marijuana, hashish, grass)
- tranquilizers and sedatives (tranquilizers, barbiturates)
- cocaine
- stimulants (amphetamine, ecstasy, ephedrine compounds)
- psychoactive substances and narcotics used simultaneously

100 ml samples of urine were collected using specialized plastic containers from persons in supervised and randomly selected groups. Preference was allotted to the determination of narcotic drugs and medicines. Use of ethyl alcohol was tested in all cases.

For persons, who were tested using the symptomatical method, blood from the elbow vein (vena ulnares), urine, or both, depending upon what substance was used (blood for determination of ethyl alcohol and its concentration; urine for determination of narcotic drugs and medicines), were taken after the evaluation of the medical examination and neurological symptoms.

2.2. Research Methods

Persons were tested referring to the approved order of physical examination for intoxication [6]: appearance of exploratory use, changes in behavior, consciousness, orientation at the point of time and place, memory, speech, walking, coordination of movement, mime, neurological and somatic condition, vegetal reactions (breathing, pulse, blood pressure), subjective and objective information about using of alcohol and other intoxicating substances, and results of additional tests and laboratory analyses.

Ethyl alcohol was identified by gas chromatography method [7; 8] using gas chromatograph Chrom-4. Conditions of chromatography: glassy column of 120 x 0.3 cm length, stationary phase chromaton N-AW HMDS 0.250-0.31 mm, liquid phase – 15% skvalan; the temperature of columns – 80°C, the temperature of injector – 60°C; speed of carrier gas nitrogen at 50 ml/min; detector katharometer, stream of detector – 80 mA;. sensibility of recording – 1:1; emission time of peaks of ethylnitrite – 20 sec., emission time of peaks of n-propylnitrite – 30 sec.

The internal standard used for quantitative determination – 2 promils (‰) n-propyl alcohol and 1‰, 2‰, 3‰, 4‰, 5‰ ethanol solution estimation curve. For conversion of ethanol water solutions the following rates were applied: for blood – 0.95, for urine~— 1.05.

Narcotic drugs have been identified by qualitative analysis, referring to the recommended methods for testing of narcotic drugs in a biological environment using screening methods in accordance with the United Nations International Drug Control Programme [9; 10].

In order to identify opiates, a 10 ml sample of urine was taken and underwent extraction by compound of chloroform and ethanol of 96° (9:1). Hydrolysis with hydrochloric and extract was performed as well. Thin-layer chromatography (THC) was performed on plate Merck , Silica gel 60 F 254", control – morphine, codeine, O6-monoacethylmorphine (MAM) 1 mg/ml in methanol.

Developing solvents included: System A – toluene acetone 96° ethanol conc.ammonia (45:45:7:3); System B – ethyl acetate methanol conc. ammonia (34:4:2). Spray reagent: Marquis and acidified potassium iodoplatinate.

For identification of cannabinoids, a 10 ml sample of urine was taken in which 2 ml 10 N kaliumhydroxid was put and heated for 15 minutes at the temperature of 50°C. Compound of hexan and ethyl acetate was used for acetyfying and extraction. Thin-layer chromatography (THC) was performed on plate Merck "Silica gel 60 F 254." Control – 9- Carboxy- THC 1 mg/ml in methanol.

Developing solvents: System A – chloroform methanol conc.ammonia (70:30:2); System B – hexan 1,4 dioxan methanol (35:10:15). Spray reagent: 0.1% aqueous solution of Fast Blue BB salt.

For determination of 1.4-benzodiazepines 10 ml of urine was alkalined by kaliumcarbonate and extracted by chloroform. Hydrolysis with hydrochloric and extract with chloroform have been performed as well. Thin-layer chromatography (THC) was performed on plate Merck "Silica gel 60 F 254." Control – benzodiazepine and benzophenons (diazepam, oxazepam, nitrazepam, clonazepam) 10 mg/ml in methanol.

Developing solvents: System A – toluene twice; System B – ethyl acetate methanol conc.ammonia (34:4:2). Spray reagent: Braton Marshall and acidified potassium iodoplatinate [11].

For determination of barbiturates 10 ml sample of urine was acetified and extracted by chloroform. Thin-layer chromatography (THC) was performed on plate Merck "Silica gel 60 F 254." Control – phenobarbital and pentobarbitalum natricum 5 mg/ml in methanol.

Developing solvents: System A – chloroform 1-butanol conc.ammonia (70:40:5); System B – ethyl acetate methanol conc.ammonia (34:4:2). Spray reagent: mercuric chloride-diphenylcarbazone. [11].

For determination of cocaine 10 ml of urine was alkalined by kaliumcarbonate and extracted by chloroform. Thin-layer chromatography (THC) was performed on plate

Merck "Silica gel 60 F 254." Control – cocaine, benzoylecgonine, ecgonine methyl ester 1µl in methanol.

Developing solvents: System A – toluene acetone 96° ethanol conc.ammonia (45:45:7:3); System B – ethyl acetate methanol conc.ammonia (34:4:2); Spray reagent: Dragendorff and acidified potassium iodoplatinate.

For identification of psychostimulants (amphetamine, extasy, ephedrine compounds), 10 ml of urine alkalined by kaliumcarbonate and extracted by chloroform. Thin-layer chromatography (THC) was performed on plate Merck "Silica gel 60 F 254." Contro – ephedrine, ephedrone, amphetamine, metamphetamine, 3.4-methylenedioxymethamphetamine (MDMA) 5 mg/ml in methanol.

Developing solvents: System A – methanol conc.ammonia (100:1,5); System B – ethyl acetate methanol conc.ammonia (34:4:2). Spray reagent: 1% Fast Black K salt in water and 1 M sodium hydroxide.

3. Research Results and Considerations

Three hundred and eighteen soldiers were tested during 2002–2004. The average term of observation was nine months. We identified 25 cases of alcohol and narcotic drug abuse (comprising 7.86 % of all samples—Figure 1).

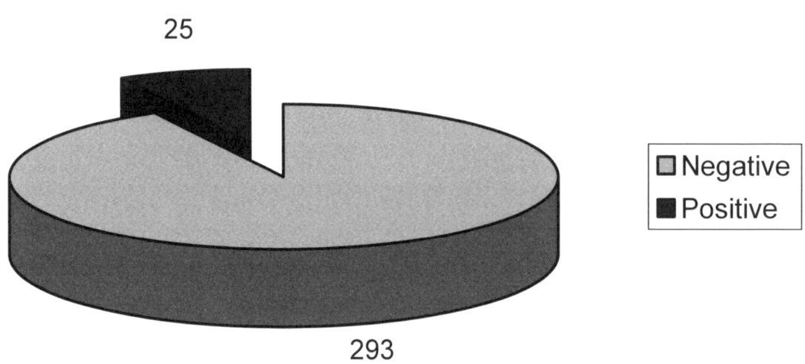

Figure 1. Positive and negative cases.

Using methods of toxicological analysis it was determined that 17 soldiers used ethyl alcohol, five soldiers used ethyl alcohol and heroine, two soldiers used amphetamine derivatives, and one soldier used carbamazepine (Figure 2).

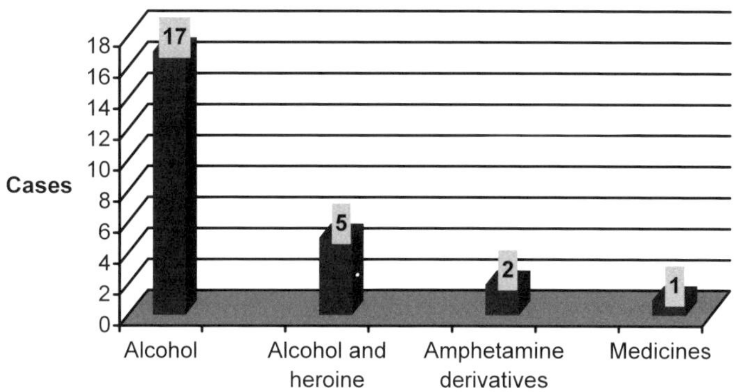

Figure 2. Used substances.

The sample was divided into three groups.The supervised group consisted of 20 persons. No instances of using ethyl alcohol and narcotic drugs were found during testing. Two hundred and seventy two soldiers were randomly selected for testing. Nine positive cases were found: 8 cases concerned ethyl alcohol and 1—using of medicine—carbamazepine (this medicine was not prescribed by doctor according to medicinal indications). Twenty-six cases were examined using the symptomatic method; 16 results were positive:

- Persons used narcotic drugs only—amphetamine and meta-0.63% all exploratory);
- Intoxication during service applied to five conscripts of initial compulsory army service. While testing, it was determined that they used ethyl alcohol and heroin simultaneously (1.57 % of all exploratory);
- Nine soldiers used ethyl alcohol only (2.83 % of all exploratory)

No cases of using of narcotic drugs were found testing soldiers of professional service. Taking all positive cases of using ethyl alcohol, sub-clinical drunkenness (0.18 – 0.35 ‰) was determined in 12 cases; slight drunkenness (0.64- 0.92 ‰) in six cases; middle drunkenness (2.19-2.41‰) in three cases and; severe drunkenness (2.53 ‰) in one case.

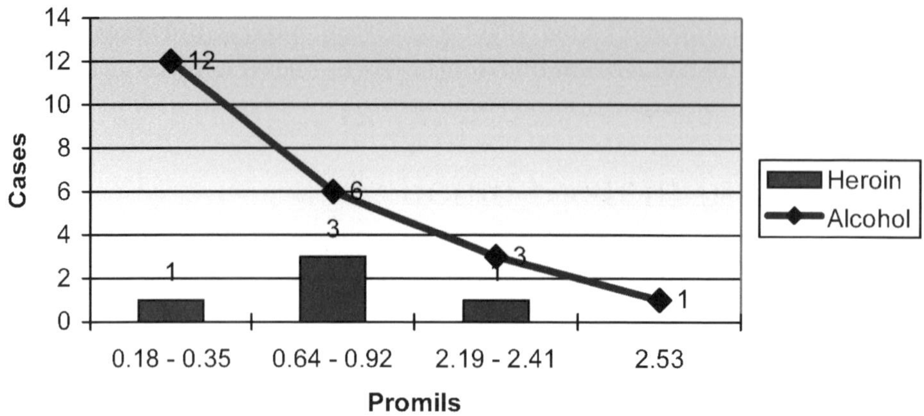

Figure 3. Concentration of ethyl alcohol.

4. Conclusions

This research study allowed us to reach the conclusion that soldiers abused ethyl alcohol, narcotic drugs, and psychoactive substances in separate cases only. The aforementioned substances were used once or twice, but they did not cause permanent mental disorder or change in behavior. As ethyl alcohol could be obtained easily, it was used most frequently (in 22 cases out of 25). There are no conditions for such harmful habits in the army.

References

[1] Dirzys D, Milasiunas R, Rivera E, Sarov J. 2001. Prevention of drugs and alcohol use and suggestions for the treatment.
Diržys D, Milašiūnas R, Rivera E, Šarov J. 2001. Narkotikų ir alkoholio vartojimo prevencija beipasiūlymai gydymui
[2] Order of Lithuanian Republic about control of drugs and psychotropic substances. 2000. Valstybes zinios; No. 4 – 113.
Lietuvos Respublikos narkotikų ir psichotropinių medžiagų kontrolės įsakymas . Valstybės žinios. 2000, Nr.4-113.
[3] Order of Health Ministry about ratification of list of drugs and psychotropic substances. 2000. Valstybes zinios., No. 4 – 113.
Sveikatos apsaugos ministerijos įsakymas dėl narkotinių ir psichotropinių medžiagų sąrašų patvirtinimo. Valstybės žinios. 2000, Nr.4-113.
[4] National Bureau of drug information. Public Health Care Service under the Ministry of Health. Drug problem in Lithuania 2002 p.Nacionalinis narkotikų informacijos biuras.Valstybinė visuomenės sveikatos priežiūros tarnyba prie SAM . Narkomanijos problema Lietuvoje 2002 p.

[5] International Statistical Classification of Diseases and Related Health Problems. 1992. Tenth Revision. Geneva: World Health Organization; 1992. 314-318p.

[6] Order of assessment of insobriety or inebriety of soldiers and civilians of National Defence System. 1999 February 5. Order of National Defence Ministry of Lithuanian Republic No. 145 of 5th Feb 1999. Karių ir civilių KAS tarnautojų neblaivumo (girtumo) ar apsvaigimo būsenos nustatymo tvarka . Lietuvos Respublikos Krašto apsaugos ministerijos 1999-02-05 įsakymas Nr.145.

[7] Zhigliavskaya OA, Novikov JA. 1982. Comparative evaluation of mathematical models of measurement of ethyl alcohol concentration in blood. Sudebno medicinskaja ekspertiza 4: 36-39p. Žigliavskaja O.A., Novikov J.A.,Sravnitelnaja ocenka matematičeskih modelej dlia opredelenija koncentraciji etilovovo spirita v krovi. Sudebno medicinskaja ekspertiza, 1982; 4: 36-39.

[8] Uspenskyj AE, Abdrashitov AX, Smirnov VM, Listvina VP. 1982. Measurement of ethanol and acetalgehyd in biological liquids. Forensic expertise 3: 45 – 48
Uspenskij A.E., Abdrašitov A.X., Smirnov V.M., Listvina V.P., Opredelenije soderžanija etanola i acetaldehida v biologičeskih židkostiah. Sudebno medicinskaja ekspertiza 1982; 3: 45-48.

[9] Recommended methods for the detection and assay of heroin, cannabinoids, cocaine, amphetamine, methamphetamine and ring-substituted amphetamine derivates in biological specimens. Manual for use by national laboratories; New York, 1995.

[10] Recommended methods for the detection and assay of barbiturates and benzodiazepines in biological specimens. Manual for use by national laboratories; New York, 1997.

[11] [Anonymous] 1986.Clarke's isolation and identification of drugs. London: The Pharmaceutical Press 168-175, 428-429, 446, 524-525, 725, 787-788, 790-791, 815-816, 883-884 p.

[12] Recommended methods for testing benzodiazepine derivatives under international control .In Manual for use by national laboratories. New York, 1988.

[13] Recommended methods for testing illicit ring-substituted amphetamine derivatives. Manual for use by national laboratories; New York, 1987.

[14] Kudriavcev JD, Ushakova MM. 1980. Qualitative gas chromatography measurement method of ethyl alcohol in small samples of blood and other biological liquids. Forensic expertise 3: 28 – 31.
Kudriavcev.J.D., Ušakova M.M.,Količestvenoje gazochromatografičeskoe opredelenije etilovogo spirita v malyh probah krovi i drugih biologičeskih židkostej . Sudebno medicinskaja ekspertiza 1980; 3: 28-31.

Public Health Preparedness Case Studies

Scientific Networking and the Global Health Network Supercourse
F. Linkov and R. LaPorte (Eds.)
IOS Press, 2006

© 2006 IOS Press. All rights reserved.

An Advanced Seismic Risk Reduction via Multinational Research Cooperation: The European Project RISK-UE

Goran S. TRENDAFILOSKI[a], Pierre MOUROUX[b] and RISK-UE Team[c]
[a] Assistant Professor, Dr. Tech.Sci., Section for Risk Disaster Management and Strategic Planning, IZIIS-Skopje, Macedonia; E-mail: goran@pluto.iziis.ukim.edu.mk
[b] Senior Expert and Project Coordinator, BRGM, Marseilles, France; email: p.mouroux@brgm.fr
[c] www.risk-ue.net

Abstract. In the last decades, besides the significant efforts being done to mitigate the destructive effects of earthquakes, there continues to be a serious threat for the world due to increased population and property density. Strong earthquakes that happened in the past revealed that modern societies can be seriously affected to the extent of decreasing, even temporarily interrupting social development with unmanageable public health consequences. Multinational cooperation and research projects in seismic risk reduction in the past and today are the consistent approach in the stated field. The European project, "An advanced approach to earthquake risk scenarios with application to different European towns (RISK-UE)," is an excellent example of such cooperation, which brought together the academic institutions from the west and the east, city officials, and the overall European scientific community in the field at one place with one main goal—to develop a general and modular methodology for creating earthquake-risk scenarios that concentrate on the distinctive features of European towns. With the participation of urban council representatives, the methodology has been applied to seven selected cities from the EU and Eastern Europe for its adaptation and validation creating European cities network for seismic risk reduction.

Keywords. RISK-UE Project, seismic risk reduction, earthquakes, GIS, risk management

Introduction

The overall preparedness of the community to cope with earthquakes should be based on relevant assessments on anticipated consequences of future events. These can be performed by: 1) extrapolation of historic losses; 2) loss-frequency analysis, and; 3) earthquake scenario analyses [1; 2].

Earthquake scenarios are the most frequently used method for loss assessment. The global indicator of their efficiency is the favorable range of the parameter cost/benefit, i.e., the predefined goals are realized with minimum financial investments.

In the past several years, pre-event creation of "earthquake-loss scenarios" utilized by different parties involved in risk evaluation and management was considered a crucial tool for systemic risk reduction. In fact, creation of a scenario makes it possible to: 1) Draw up an inventory of the elements at risk within a given town (or parts of it) using appropriate databases and computerized mapping tools such as GIS; 2) Identify weak spots within the town, i.e., areas where specific vulnerability factors could cause local damage peaks; 3) Create permanent teams with relevant knowledge for management of natural (e.g., an earthquake) or man-made disasters; 4) Set up a decision-making team involved in the creation of local (regional) and national mitigation policies; 5) Formulate risk-management plan including actions related to the rehabilitation of certain structures, the organization and preparation of the appropriate services such as healthcare system, and the entire population. The participation of emergency rescue institutions will be essential for the implementation of the risk-management plan; and, 6) Formulate an action plan ranking the steps to be taken and conduct a political-economic assessment to define the priority scale for carrying them out.

A number of approaches for constructing earthquake scenarios already exist. Especially notable is the influence exerted by the methodology integrated in the HAZUS (1999) code, launched by the U.S. Federal Emergency Management Agency (FEMA) and tailored to the characteristics of American cities. Similar programs were also proposed in the last decade. These include: 1) RADIUS (1999) for developing countries worldwide; 2) EXTREMUM (1995) developed by the Russian EMERCOM and Russian Academy of Sciences for operative forecast of consequences of destructive earthquakes with in the Council's of Europe EUR-OPA Major Hazards Agreement; 3) GEMITIS (1995) in France applied both in Nice and West Indies [3, 1, 4].

However, comprehensive programs like these did not exist in Europe. Development of such programs has proved to be an urgent requirement, particularly after the catastrophic earthquakes in Izmit, Turkey; Athens, Greece; and Chi-Chi, Taiwan. Consequently, the European project (RISK-UE), "An advanced approach to earthquake risk scenarios with application to different European towns," was launched in 1999 at the end of UN-IDNDR within the framework of the European Commission 5FP - City of Tomorrow and Cultural Heritage (EVK4-CT-2000-00014). The project was realized in the period of 2001-2004.

1. RISK-UE Project Overview and Main Objectives

The RISK-UE project is a typical example of a research effort enabling the application of the precautionary principle in urban areas. It aims to produce a decision-support tool for drawing up preventive plans of action for reducing the impacts of seismic events in European cities. RISK-UE assesses earthquake risks in diverse urban areas of Europe (taking into consideration their distinctive features) to better understand the consequences of earthquakes in terms of cost and victims, paying particular attention to historical centers, cultural heritage, public health and economic impact that directly contribute to the effective implementation of the precautionary principle at European level.

The strategic objectives of the project [5; 3; 6]:

1. *Earthquake preparedness and prevention*—To assess, within an earthquake scenario confirmed by town officials, the direct consequences in terms of cost and victims (deaths, injured, refugees), and the correlative disruption of the city's functioning over time, due to an earthquake that has a reasonable chance of occurring within the existing seismological and geological contexts. The objective is to make city officials aware of earthquake disasters in order to

improve: the level of preparedness in case of emergency situations, and land-use planning and construction practice for mitigation and prevention purposes. The criteria to be followed are the improvement of national regulations, the rate of construction renewal and the number of rehabilitation programs, in particular for strategic buildings such as health care facilities, which will have a direct impact to response activities.
2. *GIS Database*—To participate in developing urban databases structured within a Geographic Information System inventory (GIS) that will enable urban councils to carry out systematic data processing and to identify the most vulnerable areas and elements in case of a natural hazard. The database will be organized and structured around both key subjects: the phenomena (hazard) and the exposed elements (vulnerability, stakes and impacts).
3. *Web site*—To make available via a Web site the various development steps of the methodology and the results of its application to all interested parties and the general public in several towns. The Web site will ensure wider dissemination of results and, therefore, improve public awareness. Also, the Web site will be used to provide a network among towns that would enable fruitful exchange in this specific domain of earthquake-risk management.
4. *Synergy between Eastern and Western Europe*—To create a solid synergy among European countries in various regions, eastern and western Mediterranean, and particularly the Balkans (Macedonia, Romania, Bulgaria and Greece).

The scientific objectives of the project [5; 3; 6]:
1. *European methodology for earthquake scenarios*—To develop a general and modular methodology for constructing earthquake-risk scenarios that takes into consideration the distinctive features of European towns, based on the consensus of several organizations specialized in risk studies and with particular attention to historical centers, monuments, and buildings belonging to the cultural heritage of towns, economic and public health impact, and to the reaction of the population, whether individually or collectively.
2. *European distinctive features*—To gather and draw the maximum from previous experience in order to identify the distinctive features of the European context that are not considered by current methods for creating earthquake scenarios. Distinctive European urban features are particularly represented by complex building aggregates in old town centers, monuments, and historical buildings, typologies of constructions, features of transportation networks and, more generally, aspects related to specific construction techniques and the organization of towns and their response in emergency situations. The purpose of the building typology, inventory, and classification system is to set the model building classes according to the peculiarities of typical European cities, where a rich historical patrimony demands immediate protection to seismic hazard.
3. *European teams*—To set several European teams to work on developing and applying the methodology, preferably in conjunction with numerous city officials in charge. The elaborated earthquake scenarios, structured within a GIS, are to be a basis for discussion and drawing up plans of action for systematically reducing earthquake risk. The methodology is to be applied to

seven European cities: Barcelona (Spain), Bitola (Macedonia), Bucharest (Romania), Catania (Italy), Nice (France), Sofia (Bulgaria) and Thessaloniki (Greece) with different level of seismic hazard (Figure 1).

4. *Homogenize work of previous projects*—To improve and homogenize the work already undertaken in this domain as part of previous European projects, with particular emphasis on certain specific aspects, such as lifelines (healthcare facilities, transportation and telecommunications, water and energy), evaluation of indirect economic loss, direct or indirect breakdown of institutions, etc., that have been neglected during previous studies.
5. *Validation of methodology*—To validate this methodology through a network of experts.
6. *Standards*—To produce a standard manual for assessing earthquake risk in urban areas for use not only by all European countries, but also by countries with a habitat and lifestyle similar to European conditions, particularly those in the Mediterranean region.

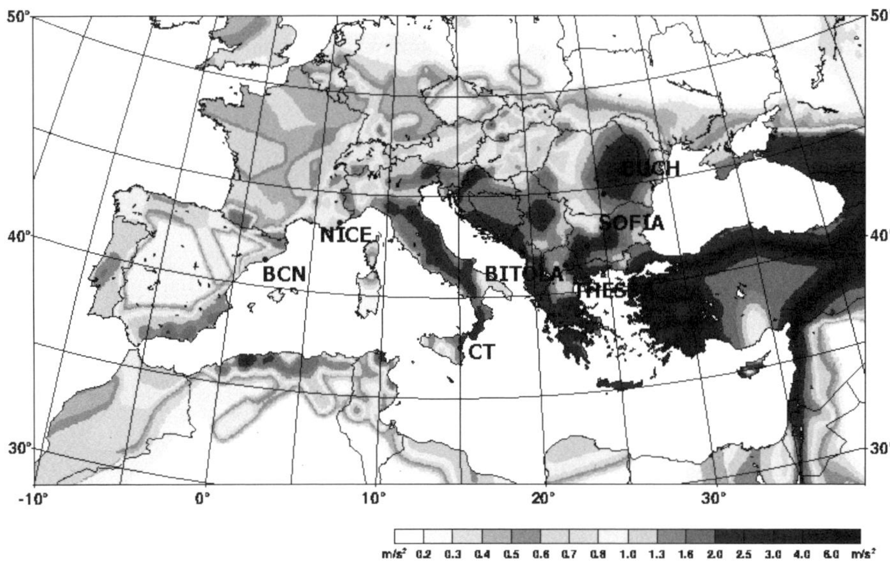

Figure 1. RISK-UE Cities and GSHAP 475 yrs seismic hazard map (source: GSHAP Project http://seismo.ethz.ch /gshap/)

2. Organization of the RISK-UE project

The RISK-UE project is organized in (Figure 2) [3]:

- Seven methodological work packages (WP1-WP7): 1) **WP1**—European distinctive features, GIS inventory database and building typology; 2) **WP2**—Seismic hazard assessment at both regional and local level; 3) **WP3**—Urban system analysis, identification of weak points in towns under normal conditions, during crisis, and recovery periods; 4) **WP4**—Vulnerability assessment of current buildings; 5) **WP5**—Vulnerability assessment of old town centers, historical monuments, and buildings; 6) **WP6**—Vulnerability assessment of lifeline facilities and essential structures; 7) **WP7**—Seismic risk scenarios.

- Seven work packages (WP8-WP14) related to practical application of the RISK-UE methodology in the selected cities.
- Two work packages regarding coordination and management of the project (WP0) and dissemination of results (WP15).

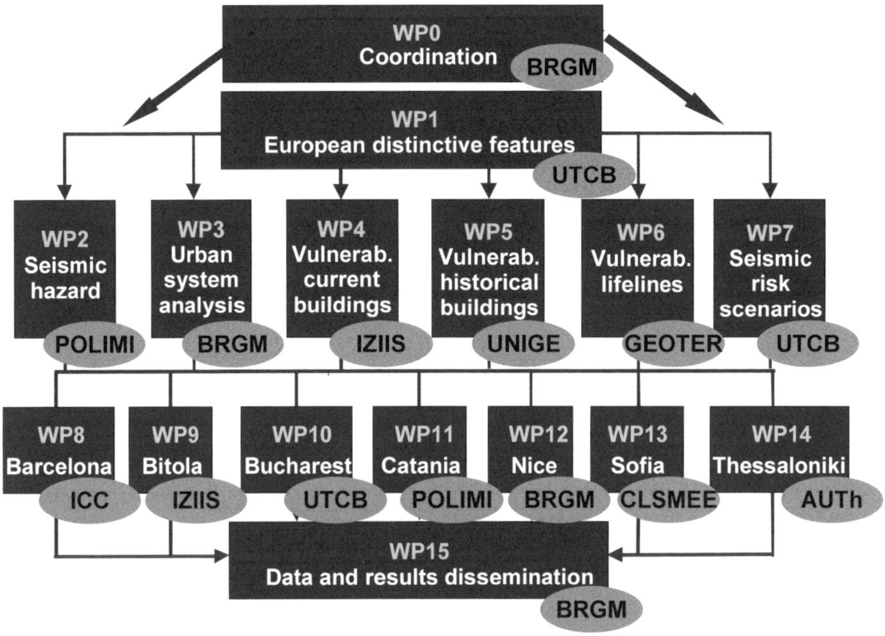

Figure 2. RISK-UE Project organization [6]

The consortium of 10 university institutions and authorities of seven European cities (Table 1) took part in the performance of the project, thus, creating a network of different professionals in the field for achieving of the project's objectives.

3. Project achievements

Throughout the project's performance the following achievements will be highlighted [3]:

- Creation of a basic tool for the decision makers responsible for risk management by means of implementing an earthquake scenario at city scale (Figure 3).
- Completion of a European manual for creation of earthquake scenarios taking into account the European distinctive features.

Table 1. RISK-UE project consortium [3]

Institution/ City	Name
BRGM	Land Use Planning and Natural Risks, Marseilles, FR
GEOTER	Geology, Tectonics, Environment and Risks, Roquevaire, FR
POLIMI	Politecnico di Milano, Structural Engineering Dpt, I
UNIGE	Univ. of Genoa, Struct. and Geotech. Dpt, I
UTCB	Tech. Univ. of Bucharest, Civil Eng. Dpt, RO
ICC	Cartographic Institute of Catalunya, Barcelona, E
AUTH	Aristotle University of Thessaloniki, GR
IZIIS	Inst. of Earthquake Eng. and Eng. Seismology, Skopje, MK
CLSMEE	Lab for Seismic Mechanics and Earth. Eng., Sofia, BG
CIMNE	Int. Center for Numerical Methods in Eng., Barcelona, E
Cities	Ajuntament de Barcelona, E
	Comune di Catania, I
	Ministry of Public Works, Transport and Housing, RO
	Ville de Nice, FR
	Municipality of Sofia, BG
	Town of Bitola, MK
	Organization of Thessaloniki, GR

- Assessment of a plausible earthquake, validated by city representatives, in terms of direct and indirect damages (cost, victims), with particular attention to old town centers, cultural heritage, the public health and economic impact.
- Development of a database regarding natural phenomena, elements at risk, and their vulnerability.
- Establishment of short, medium, and long term Management Plans and Action Plans.
- Creation of inter-city Euro-Mediterranean synergy enabling permanent exchanges with regard to the problems posed by natural disasters.

The RISK-UE project involved city officials and local stakeholders, thus creating the foundation for local governance of risk and environmental management. Based on both the voluntary participation of cities and the mobilization of experts and field players, this project has contributed to: 1) Establishing a risk-reduction and environmental protection policy to be integrated into the urban planning and development plans, thus promoting an appropriate legislative and standards framework; 2) Adapting community organizations by developing municipal structures for risk management; 3) Increasing awareness of local officials in terms of risk reduction and crisis management; 4) Spreading information and increasing the awareness of the public in order to prepare the most exposed sectors for future crisis situations; 5) strengthening scientific and technical cooperation; 6) Launching a European initiative for open exchange among local authorities, mindful of the differences; and, 7) Establishing a common-support framework in the event of natural disasters.

Another contribution of this project to the social objectives of the European community is its political dimension i.e., the involvement and active participation of EU candidate countries (Bulgaria, Romania and Macedonia).

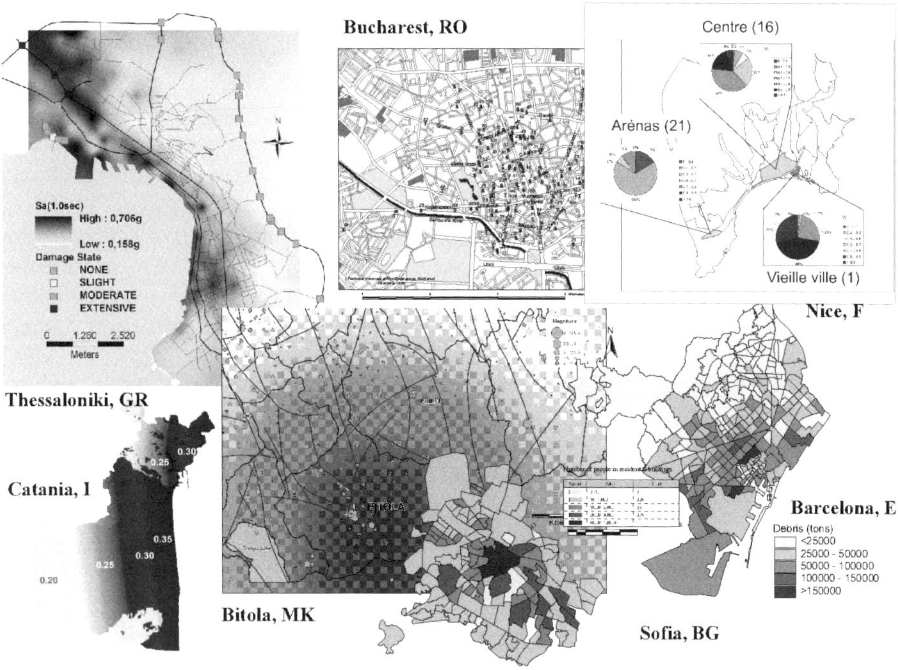

Figure 3. Earthquake-risk scenario maps for RISK-UE cities [2]

The prospective disaster scenarios developed in the project closely related to the main urban actors (mayors, economic leaders, municipal departments, public and private services, healthcare officials, non governmental organizations {NGOs}, the media). The project's strategy is adapted to the European conditions, gaining local appropriation of preventive actions as part of the sustainable development of cities, protecting cultural heritage, and forming the basis for improved safety of the population. The RISK-UE is based on the same principles as Local Agenda 21, thus, making it possible to improve actual European policies concerning local sustainable development and environmental management.

The dissemination of the project results has been achieved using different tools such as: 1) Urban GIS and mapping; 2) Videotapes, TV's and interviews; 3) Web site <http://www.risk-ue.net>; 4) Plenary and working meetings; 5) Special sessions at: 12th ECEE, London 2001; 3rd SCEE, Malaga, Spain; 13th WCEE, Vancouver 2004 and, 6) Nice Final Conference (April 2004) organized with participation of the main actors.

4. Future activities

The RISK-UE consortium, as a whole or its individual members, will continue previously established cooperation through creation of a special scientific network integrated by member cities and open to other interested (or associate) cities prone to earthquake disasters. Each associate city will involve representatives of the municipality, scientific institutions, and emergency services.

The consortium intends to exploit all the work realized inside the methodology and developed in the application for the seven cities, as follows [2; 7]: 1) Continuous improvements of the methodology by keeping a permanent European working group, in which any other representatives relevant institutions could be included; 2) Continuous database updating by all cities; 3) Continuous implementation and improvement of the risk management and action plans. The action plan will emphasize the priorities regarding land-use, retrofitting of constructions, emergency/assistance and relief, awareness (information and training), depending upon motivation and appropriation of seismic risk by different actors of the city and financing possibilities of any kinds; 4) From the network established by these seven cities, the settlement, as soon as possible, of a network with other Euro-Mediterranean Cities such as Algiers, Rabat, Tunis, Cairo, Nicosia, Ljubljana.

Acknowledgments

This paper presents an overview of the global European research activities for developing an Advanced Approach to Earthquake Risk Scenarios with Application to Different European Towns (project RISK-UE), partially financed by the European Commission (Contract No. EVK4-CT-2000-00014). The authors extend their most sincere gratitude for the financial support of the project and the collaboration of all members of the RISK-UE consortium during the course of project performance.

References

[1] Trendafiloski G. 2003. GIS-oriented method for elaboration of probabilistic earthquake scenarios. [doctoral dissertation]. Macedonia:IZIIS-Skopje.

[2] Trendafiloski G, Mirakovski LJ, Cvetkovska B. 2003 May. Assessment of possible regional consequences, Invited lecture, SEEBRIG Military-Civil Emergency Planning Engineering Seminar, Plovdiv, Bulgaria.

[3] Mouroux, P. (coord.) 2004. RISK-UE An advanced approach to earthquake risk scenarios with applications to different European towns. Final report, BRGM, France.

[4] Trendafiloski G, Mirakovski LJ, Cvetkovska B. 2003 May. Computer models for damage assessment after seismic impact, Invited lecture, SEEBRIG Military-Civil Emergency Planning Engineering Seminar, Plovdiv, Bulgaria.

[5] Milutinovic Z, Trendafiloski G, Mouroux P. and RISK-UE Team. 2003 June. Project RISK-UE. Invited lecture. Council of Europe EUR-OPA MHA International Workshop on Seismic Safety and Emergency Management of Essential Facilities, SEMEF-2003, Ohrid, Republic of Macedonia.

[6] Mouroux, P, Bertrand E Bour M, Le Brun B, Depinois S,. Masure P, and RISK-UE Team. 2004. The European RISK-UE Project: An advanced approach to earthquake risk scenarios; Vancouver, Canada. Proceed.of 13WCEE, pap. ref 3329..

[7] RISK-UE (2001-2004). Available from: http://www.risk-ue.net

Scientific Networking and the Global Health Network Supercourse
F. Linkov and R. LaPorte (Eds.)
IOS Press, 2006

© 2006 IOS Press. All rights reserved.

Emergency Preparedness in the Baltics

Dinas VAITKAITIS
Kaunas University of Medicine, Kaunas, Lithuania (dinas@kmu.lt)

Abstract. The aim of this paper is to describe the risks, emergency management, emergency services, and education and training in the Baltic countries. Disaster management is divided into three levels in Lithuania: national, regional, and local. Medical disaster preparedness organization in Latvia is divided into three levels: state—Ministry of Health, Centre of Emergency and Disaster medicine, operative reserves; district—responsibility for preparedness; local—ambulance service and hospitals. Facing the same threats, our very similar countries have similar problems in the area of emergency preparedness. Solutions should also be similar.

Keywords. Baltics, emergency preparedness, public health

Introduction

The aim of this paper is to describe risks, emergency management, emergency services and education and training in Baltic countries. The Baltic region consists of three European Union (EU) countries located on the east coast of the Baltic Sea: Estonia, Latvia, and Lithuania. They have a similar history, geography, economy, and social (Table 1). There are approximately more than 7.5 million people living in these countries in a territory of 175 thousands square kilometers. Lithuania is the largest of the Baltic countries in terms of size and population.

Table 1. The Baltic countries: geography, demographics and economy

	Lithuania	Latvia	Estonia
Geographic size, Thousands km2	65,3	64,5	45,0
Population size, Millions	3.6	2.44	1.5
GDP, Billions	$45.23	$26.53	$19.23
GDP, per person	$12,500	$11,500	$14.300
% of GDP spent on healthcare	5.2	5.1	5.1

Latvia and Estonia might be considered countries composed of capital cities, as the majority of their populations are located in major cities: two-thirds of the inhabitants are living in one-third of the overall territory.

1. Natural and Man-made Disaster Risks Similar in Baltic Countries

The risk of natural disasters is relatively low in the Baltic countries. Spring floods are common in delta areas. Forests fires are related to precipitation in that the risk increases in drought situations. Stormy winds are common. The risk of man-made disasters includes nuclear power plants, transportation of toxic substances, automobile accidents, and terrorism.

The risk of man-made disasters is similar in all three countries. For example, the working Chernobyl-type reactor at the Ignalina Nuclear Power Plant is a high risk facility. In 1994, there was a serious accident causing acute radiation injuries, and even death, in Tammiku, Estonia: a victim found a container with Cs-137 in a dump, and took it home; three more persons were irradiated.

In addition to nuclear reactors, tons of dangerous chemical substances vulnerable to spills are transported by roads and railways. Several accidents involving toxic substances happen every year in all three countries, the most notable of these happened in Lithuania in 1989. The accident was billed as the world's largest disaster involving ammonia where more than 7,000 tons were spilled. The official count of deaths and injuries was: seven dead, 67 suffered from intoxication. The spill caused extensive damage to the environment.

The other man-made disasters are motor vehicle accidents. They are among the main killers in the Baltic countries—the mortality rate in Latvia and Lithuania is one of the highest in the EU. In addition, terrorism is a main risk factor as seen in the bombing of the Riga shopping center in August 2000, where one person was killed and 30 were injured.

2. Structure of Emergency Management

Disaster management is divided into three levels in Lithuania: national, regional and local. At the national level, the government institutions are responsible for preparedness and management in the case of disasters, which includes the emergency management commission and emergency management center at the government and ministries. Their responsibilities are defined by the law.

At the regional level, authorities are responsible for the emergency planning and establishing emergency management centers. At municipalities the local authorities are responsible for the same matters.

3. Emergency Medical Services in the Baltic Countries

3.1 Overview

Emergency management in the health care system follows the same principles: government level, regional administration, and municipalities. At the local level, what is termed "municipality doctors' institutions" are responsible for emergency planning. Hospitals are divided into three levels: university hospitals, county hospitals and municipality hospitals.

3.2 Lithuania

In case of an accident, emergency services including the fire-rescue service, police, and health services are activated. The fire-rescue service is responsible for rescue work, protection of persons and property from damage, first aid at scene, and transportation of the injured people. The police play a coordinating role at the scene. They protect persons and properties, and also control traffic, register victims and bystanders.

Paramedics and health care personnel triage victims and ferry them by ambulance from the site of the accident to hospitals. There are 63 ambulance stations with a total of 265 teams, who operate during the day, and 210 teams who cover the nightshift.

In 2002, the Concept of Ambulance Services was adopted by Ministry of Health of Lithuania. According to this new regulation, there will be changes in the entire emergency system: re-equipment of ambulances, establishment of paramedic services, and re-arranging emergency departments at hospitals. All three services will be responsible for the planning and exchange of information. Other services may respond to emergencies when needed such as search and rescue and air search and rescue (with seven heavy helicopters available.

3.3 Estonia

In Estonia, emergency medical services (EMS) are decentralized. There is no governmental body regulating the EMS activities. The Union of Estonian Emergency Medical Services is a non-profit organization for EMS professionals (physicians, nurses and EMTs), devoted to quality emergency medical care by providing training and guidance to its members, and consulting the government in questions vital for the development of EMS, etc. There are 87 EMS teams in 46 geographic locations. EMS teams in Estonia include six animobiles with anaesthesiologist on board (based in Tallinn and Tartu), 40 physician-manned ambulances (generalists with training in EM), and 41 teams lead by a specialist nurse. All EMS teams consist of three members: a doctor or a specialised nurse, a nurse, and a driver trained as EMT.

3.4 Latvia

The Medical disaster preparedness organization in Latvia is divided into three levels: 1) State—Ministry of Health, Centre of Emergency and Disaster medicine, Operative reserves; 2) District—Responsible for preparedness; 3) Local—Ambulance service and hospitals. There are 52 ambulance stations in 70 localization points. 173 teams work during the day, and 143 during the night. There two types of EMS teams: Intensive therapy (IT) teams and emergency (EM) teams. IT teams consists of a three members, a doctor, doctor's assistant, and driver, divided among 53 daytime teams and 36 night teams. EM teams consist of a doctor's assistant and a driver who are assigned to 123 daytime teams and 110 night-time teams. Currently, there is legislation that requires two medical educated persons on each team.

4. Education and Training in the Baltic Countries

In Lithuania, medicine is taught at three medical universities and five health faculties at colleges (nursing schools). Currently, we are working on postgraduate training and exercises. A year ago, an emergency medicine research program was approved by Kaunas University of Medicine, and we are working on implementing trauma training. In 2003, we started advanced cardiac life support training. There is strong cooperation with military services in carrying out emergency exercises. We use the EmergoTrain System for table-top exercises and for conducting large-scale exercises.

In Estonia, undergraduate medical education for physicians is provided at the Tartu University. Nursing programs are offered at three schools: Tallinn, Tartu, and Kohtla-

Järve. There is a residency program at Tartu University Hospital for EM physicians, and EMTs are trained at Tallinn and Tartu EMS, Center of Disaster Medicine of Tallinn Mustamäe Hospital.

5. Conclusion

Facing the same threats, our very similar countries have similar problems in the area of emergency preparedness. Solutions should be similar.

Scientific Networking and the Global Health Network Supercourse
F. Linkov and R. LaPorte (Eds.)
IOS Press, 2006

© 2006 IOS Press. All rights reserved.

State Fire and Rescue Service: Our View of Scientific Networking and Global Health Supercourse

Vladas MIRECKAS, M.D.
Fire and Rescue Department under the Ministry of Interior

Abstract. This paper briefly reviews the activities of the fire and rescue department, the structure of the fire and rescue forces, as well as their role in responding to emergencies in Lithuania. It also provides legal basics of the fire and rescue department, and the aforementioned forces in the field of civil protection. Finally, this article provides an opinion about the principle activities of the national health care system in preparing for emergencies, and outlines our views on the importance of scientific networking for the rescue services.

Keywords. Fire and rescue services, emergency mitigation, Lithuania

Introduction

Lithuanian Fire and Rescue forces are divided into two main types: state fire and rescue service, and municipal fire brigades. The State Fire and Rescue Service consists of the Fire and Rescue Department under the Ministry of Interior, 55 fire and rescue services, which are located in the main cities across the country, Firefighters Training School, and the Fire Research Center. Approximately 4,800 firemen are employed at state-owned fire and rescue services, and another 2,100 at municipal fire brigades.

1. Civil Protection and Rescue System

In accordance with the law on civil protection of the Republic of Lithuania, the State Fire and Rescue Service shall be a constituent part of the Civil Protection and Rescue System, which is in the state of constant readiness organizing the extinguishment of fires, extinguishing fires, carrying out rescue operations, as well as state fire prevention supervision.

The State Fire Prevention and Rescue Service, municipal fire prevention services, departmental fire prevention forces, and voluntary firemen units taking part in emergency response operation shall be subordinate, regardless of their institutional dependence, to the commander of rescue operations appointed, in the prescribed manner, by the director of the Fire and Rescue Department under the Ministry of the Interior (hereinafter, referenced as the Fire and Rescue Department).

The Fire and Rescue Department shall be a constituent part of the civil protection and rescue system directing the activities of the civil protection and rescue system, organizing the prevention of emergencies, coordinating the activities of state institutions and economic entities in the sphere of civil protection and planning the national preparedness for the implementation of civil protection tasks in the event of an emergency, in time of peace, and in wartime.

2. Implementation

When implementing the assigned tasks, the Fire and Rescue Department shall plan measures in order to ensure the transfer of state institutions, economic entities, and the population from ordinary working (living) conditions to the state of emergency with the smallest possible losses, maintenance of order, preservation of human life, health and property and protection of the environment against the effects of the emergency.

3. Control the Activities of the Civil Protection and Rescue System

The Civil Protection and Rescue System organizes the preparedness of state institutions and the community for practical actions in case of an emergency; coordinates the preparation of plans of civil protection preparedness for an emergency in counties and municipalities, and; approves county plans of civil protection preparedness for an emergency.

4. Purpose of the Service

The purpose of the service shall be to organize the extinguishment of fires, extinguish fires, carry out rescue operations, respond to natural disasters, chemical and radiation incidents, as well as provide first medical aid to the victims.

The law mentioned above also states that medical care and public health care institutions must be prepared to organize their activities in the event of an emergency in accordance with a plan of civil protection preparedness of the agency for an emergency, which has been prepared in advance. The plan shall be developed according to the recommendations of the Ministry of Health.

In view of the stated circumstances the Fire and Rescue Department is interested in establishing an efficient and science-based system for the activities of the personnel and public healthcare institutions in the event of emergencies. Our point is that the system in question should be based on the following principles: (a) Constant readiness (standby), (b) Advance planning of actions, (c) Modular training, (d) Inter-institutional and international cooperation, (e) Application of scientific knowledge.

5. Conclusion

As a former Soviet Union country, Lithuania has much to accomplish in order to establish such a system, which seems to be a difficult task without appropriate knowledge. We are convinced that the establishment of the scientific research data network in the field of emergency prevention and mitigation is an excellent and timely instrument for the dissemination of such information. This system will enable the countries participating in the project to advance more rapidly in their preparedness activities for the emergency response.

F. Linkov and R. LaPorte (Eds.)
IOS Press, 2006
© 2006 IOS Press. All rights reserved.

Concepts and Terms Used by the Government of Estonia in Health Care Preparedness for Disasters

Ursel KEDARS, M.D
Ministry of Social Affairs of Estonia, Public Service Academy

Abstract. This article gives an overview of terms, definitions, concepts, ideology, the methodological basis, organization, responsibilities, and the legal framework currently held by the Estonian government in health care preparedness for disasters. The Crisis Committee of the Government of Estonia, as well as the Ministry of Social Affairs, has a central role in organizing activities aimed at prevention and limitation of the effects of a disaster on public health.

Keyword. Crisis Committee of the Government of Estonia, disaster, public health, preparedness

1. Definition of Disaster and Area

From the medical point of view, disasters (i.e. emergencies) in Estonia are characterized by an acute and unforeseen imbalance between the capacity and resources of the health care system and the needs of the victims, or the people whose health is threatened, over a certain period of time. During the initial phase of the event it is also essential to take into account the number of victims.

2. Ideology and Management

Disaster preparedness of the health care system is based on disaster medicine ideology. The disaster medicine ideology encompasses emergency medicine, public health techniques, and common emergency preparedness for any accident. The idea incorporates health care providers like hospitals and emergency medical services (EMS, ambulances), prevention of infectious diseases, public health, and the medical stockpile. It also integrates the non-medical resources, capabilities, and services including fire and rescue services etc., in close cooperation among central and local authorities, in accordance to a strategy aimed at prevention and limitation of the effects of a disaster on the health of the populations concerned. The strategy, which is aimed at limitation of the effects of a disaster on health, is reached, in particular, through saving the lives of and limiting the development of further affects on as many victims as possible.

In practice, the theory of Incident Management System (IMS) and its modification Incident Medical Management System (IMMS) on-scene methodologies are implemented and used in case of common accidents. In principle, the advanced solution of IMS is taken into consideration in health care disaster and emergency management methodologies. According to the IMS methodology an On-scene Operation Commander (OSOC) manages the whole on-scene rescue operation, while the On-Scene Medical Commander manages the medical sub-operation. Temporary headquarters make it possible to build up immediate on-scene cooperation between commanders of different services.

In practice, the OSOC fulfils the tasks of overall Rescue Work Coordinator. In case of an emergency, a high level rescue operations manager will take over the tasks of the Rescue Work Coordinator, but the Crisis Management Team or even the Crisis Committee may also be summoned to take responsibility for the activities. In special cases of health care emergencies the Emergency Medical Coordinator will organize the medical work and take the functions of all-over medical command. The Emergency Medical Coordinator will work in cooperation with the Ministry of Social Affairs, On-scene Medical Commander, and other members of the headquarters of the rescue work. Depending upon the dimension of the accident—emergency, pre-disaster or disaster—different levels of government will be involved in managing the rescue operation.

3. Government Agencies

The Ministry of Social Affairs is responsible for disaster preparedness of the health care system. Two government agencies in the administrative area of the Ministry of Social Affairs have partial responsibilities for disaster preparedness in health care. The first of these agencies is the Health Care Board that deals with the organization of health care and the Health Protection Inspectorate with communicable diseases and public health issues. Although the Ambulance Service (EMS) is operated by the Ministry of Social Affairs and the Health Care Board, the command center (EMS dispatch), is operated by the non-medical Rescue Board under the Ministry of Internal Affairs in accordance with the directives of the Ministry of Social Affairs.

Second, the crisis (management) committee of the Government of the Republic is the principal structure summoned by the government in order to develop a crisis management policy and make proposals for coordinating and directing the activities of the ministries in crisis management. The members of the Crisis Management Committee of the government are the ministers and secretary generals including the Minister and Secretary General of Social Affairs. The important satellite structure to assist the main committee is a Crisis Management Team, summoned by the Crisis (management) Committee of the Government, which closely cooperates with the crisis management committees of counties.

Third, the Ministry of Internal Affairs provides professional support to the Crisis Management Committee and is responsible for general crisis prevention and crisis management, and organizes civil defence and rescue operations (excluding ambulance service). In the context of crisis management, the Ministry of Internal Affairs has to prepare the National Crisis Management Plan in close cooperation with other ministries and submit the plan to the Government of Estonia for approval. The health issues in National Crisis Management Plan are provided and prepared by Ministry of Social Affairs.

4. Responsibilities of the Ministry of Social Affairs

Through the Emergency Preparedness Act, effective January 1, 2001, the Ministry of Social Affairs in its area of government (health care organization etc.) has to:

- form a crisis management team of Ministry of Social Affairs;
- perform risk assessment in order to identify the emergencies that might occur,
- and organize crisis management training in its area of government (health care organization, health care services provision, public health etc.);
- prepare a crisis management plan for health care and ensure the implementation thereof in an emergency;
- plan the use of health care resources necessary for responding to an emergency;
- perform an assessment in its area of government in order to identify the
- resources that are necessary for participation in international crisis management;
- determine the crisis management duties of agencies and inspectorates in the
- area of government of the ministry;
- form a structural unit in the ministry to be responsible for the general
- organization of crisis management in the area of government of the Ministry of Social Affairs;
- direct the response to an emergency in its area of government i.e., health care
- emergency and medical field operations.

By law, the vitally important areas of government that the Ministry of Social Affairs has to administer are the Organization of Health Care (including hospitals, transportation of patients and ambulance service, primary health care, prevention of spread of communicable diseases, health protection, other medical management and guidance), assistance to refugees and the evacuated, but also health and social insurance, social welfare, provision of psycho-social help, and thereby, participation in crisis management.

5. Health Care Emergency

Health care organization in emergencies and of emergency preparedness is established by the Regulation of the Government of the Republic (Regulation No. 222 of 10 July 2002). The regulation sets up the definitions of Health Care Emergency (i.e., disaster) and Health Care Emergency Medical Coordinator. In case of health care emergency the responsibility for provision response is the Minister of Social Affairs. The functions of preparation for and coordination of solving the health care emergency does have a crises management team of Ministry of Social Affairs in collaboration with Health Care Board. In case of a health care emergency, the minister appoints the Emergency Medical Coordinator.

In case of health care emergency, enormous numbers of victims needing medical aid are expected to appear. The possible reasons for injuries or illnesses would be outbreak of communicable diseases, man-made large-scale accidents (including terrorism), natural disasters, radiation exposure, and intoxicants. Note that Estonia transforms its governmental preparedness for emergencies in the field of health care in close collaboration with other member-states of European Union.

6. Legal framework, terms and definitions

Terms and definitions used in Estonian health care organization in the field of disaster preparedness based on legal framework are as the following:

	Legal framework
Emergency (disaster) – event or a chain of events which endangers the life and health of persons, national security, significantly damages the environment or causes extensive economic damage, and responding to which requires the coordinated action of the Government of the Republic, government agencies, and local governments.	EPA
Health care emergency – event or chain of events which endangers the life and health of people and has brought out or potentially may cause life threatening illness, injury or intoxication of many people, and there is an ineluctable need for the simultaneous concerted action of Government of the Republic, Ministry of Social Affairs and its Health Care Board, Health Protection Inspectorate, regional hospitals and health services providers.	Gov Regulation No 222 of 10 July 2002
International emergency – event or a chain of events in a foreign state which endangers international security, and which affects the Republic of Estonia through international obligations or through an international request for assistance.	EPA
Emergency situation (state of emergency) - declared on the condition that it is necessary to implement extensive measures, including rescue work, promptly in the event of a natural disaster or catastrophe, or in order to prevent the spread of an infectious disease, and that it is not possible to eliminate the danger or assist the victims without implementation of the measures provided for in this Act. Declared to resolve an emergency. The Government of the Republic declares an emergency situation throughout the state or in a part in the event of a natural disaster or catastrophe or in order to prevent the spread of an infectious disease.	ESA (NSA)
Crisis management –the state system of measures to be prepared and used by state agencies in cooperation with local governments, undertakings, non-profit associations and foundations involved in crisis management in order to ensure public safety in emergencies.	EPA
International crisis management –the system of international measures for responding to international emergencies which involves international cooperation and cooperation with international organizations.	EPA
Crisis management team – means a permanent team formed in order to coordinate exchange of information, use of resources and cooperation between agencies and to analyze the situation, and the designated staff, organization of work and training of which enable the team to work 24 hours in the case of emergency. In case of a health care emergency the main importance does have the crises management team (CMT) of Ministry of Social Affairs, but also CMT of County Governments and CMT of the Government of Republic of Estonia have important roles.	EPA
Emergency medical coordinator – in case of health care emergency all health care services providers must fulfil commandments of minister of social affairs and	Gov Regulation

emergency medical coordinator. Executive arm of regional hospital has to take orders of emergency medical coordinator. In case of health care emergency minister appoints the emergency medical coordinator.	No 222 of 10 July 2002
Rescue work coordinator – a rescue leader who is in charge of all fire extinguishing and rescue teams and additional forces arriving on the scene of an accident, including ambulance vehicles and teams (i.e., OSOC).	RA
(*Temporary*) **Headquarters** – temporary management structure which staff members should be medical leader, police leader, rescue leader; established by the rescue work coordinator.	MoI Regulation No.4 of 23 Feb. 2000
Natural disaster – an event which causes severe damage, including an unexpected event which causes severe damage, which is caused by the forces of nature and which endangers life, health, or the natural or manufacturing environment.	ESA
Catastrophe – an unexpected event which causes severe damage, which endangers life, health, or the natural or manufacturing environment and which consists of one of the following: • the chemical, radioactive or other contamination of a locality; • an industrial accident, including accidents at power stations and mines, and accidents involving gas pipes, telecommunications networks, public utilities networks and power networks; • an extensive fire or explosion; • an extensive accident involving a ship, aircraft, train or other means of transport; • other extensive accidents.	ESA
Spread of infectious disease – a life-threatening infectious disease spreading as an epidemic, including as an infectious animal disease, the consequence of which is a widespread outbreak of illness and against which it is necessary to carry out extensive mitigation; the carrying into Estonia of an extremely dangerous infectious disease subject to international notification which is capable of spreading as an epidemic, and in order to prevent the spread of which and to avoid a widespread outbreak of illness it is necessary to carry out immediate and extensive mitigation.	ESA
National stockpiles – the aggregate of tangible resources and technical means (hereinafter resources), which ensure the security and independence of the Republic of Estonia, and which are used in an emergency or during a state of war. Definitely the stockpile is the aggregate of resources at the disposal of the state which is used to ensure the survival of the population in an emergency (i.e., disaster). The procedures for the replenishment, storage and use of the security stockpile of medicinal products and medical supplies in the national stockpiles sets out the Regulation of the Government.	NSA HSOA

7. Abbreviations

CMT Crisis Management Team

EMS Emergency Medical Services i.e., Ambulance Service

EPA	Emergency Preparedness Act
ESA	Emergency Situation Act
HSOA	Health Services Organization Act
IMS	Incident Management System
IMMS	Incident Medical Management System
MoI	Ministry of Internal Affairs
MoS	Ministry of Social Affairs
NSA	National Stockpiles Act
OSOC	On-Scene Operation Commander
RA	Rescue Act

References

[1] Regulation of the Government of the Republic, Regulation No 222 of 10 July 2002
[2] Emergency Preparedness Act, entered into force 1 January 2001(amended)
[3] Emergency Situation Act, entered into force 16 February 1996 (amended)
[4] Rescue Act, entered into force on 24 April 1994 (amended)
[5] National Stockpiles Act, entered into force 30 December 1994 (amended)
[6] Health Services Organization Act, entered into force 9. May 2001 (amended)
[7] Public Health Act, entered into force 21 July 1995 (amended)
[8] Katastroofi tagajärgede likvideerimise metoodika. Teenistuste koostöö- ja valmisolekuplaan. Metoodiline juhend, Tallinn 2003/2004

Scientific Networking and the Global Health Network Supercourse
F. Linkov and R. LaPorte (Eds.)
IOS Press, 2006
© 2006 IOS Press. All rights reserved.

Methodology for Eliminating the Consequences of a Disaster in Estonia

Raul ADLAS, MD[1]
Tallinn Emergency Medical Service

Abstract. Methodology for eliminating the consequences of a disaster presents methodological guidelines for the cooperation of operational services and other agencies at the disaster site to eliminate the consequences of major accidents or disasters. This plan has been developed by the specialists of the Estonian Emergency Medical Services, The Police Board, The Rescue Board, The Defense League, and North- Estonian Regional Hospital.

Keywords. Crisis Management, disaster, preparedness, rescue operations

Introduction

This plan was developed with the assistance of specialists from the Kingdom of Sweden; the disaster medicine plan of the county of Stockholm was used as a basic model. This plan has been successfully implemented in many theoretical and practical exercises, and at several complex field exercises. For the purpose of this plan, the following terms are defined in a shortened version:

- Disaster—an unexpected event, which causes severe damage and which endangers life, health, or the natural or manufacturing environment. An incident, which results in the need of so significant resources in comparison to the existing ones that it is impossible to maintain normal standards of life even after taking extreme measures [1].
- Public Health Emergency—an event or a chain of events, which endangers national security, the life and health of persons, significantly damages the environment or causes extensive economic damage, and responding to which requires the coordinated action of the Government, government agencies and local governments [2].
- SOS Center—alarm center, a structural unit of the Rescue Board, which receives incoming notices of accidents, dispatches ambulance crews, fire extinguishing and rescue teams to the injury site and organizes operational communications and information processing. The SOS center receives emergency calls by common telephone number 112.
- Injury Area —an area where a disaster has occurred, fire extinguishing and rescue operations are being administered and the structures eliminating the

[1] Raul Adlas, MD is the Head Doctor of Tallinn Emergency Medical Service, head of the group of specialists, working out the mentioned plan.

consequences of the disaster are located. For acting in the injury area, restraints are imposed on persons.

- Injury Site—an area where damage has arisen and/or where victims and valuable evidence for clarifying the occurred event are located.
- Management Post—a site in the injury area where a rescuer directing the fire fighting and rescue operations, a medical team and a police officer and a communications center are located. The work of the management post is directed by the injury site rescue coordinator.
- Check Point—a site where all vehicles participating in the rescue operation shall arrive and from where all other traffic shall be stopped and redirected. (The checkpoint shall not be mistaken for an ordinary roadblock, several of which can be erected and which might not be an appropriate assembly point for the reserve).
- Rescue Work Coordinator—the chief of rescue operations, a rescuer directing the rescue operations from the management post, coordinating the cooperation of the medical officer and senior police and the whole activity in the injury area.
- Senior Police—an executive police officer directing the action of the police in the injury area.
- Medical officer—a senior medic with special training in disaster medicine who is in charge of the medical side of the rescue operations at the management post.
- Injury Site Doctor—a physician/nurse with special training in disaster medicine, in charge of the work of medical personnel at the injury site.

1. Objectives of the Cooperation and Preparedness Plan

The Cooperation and Preparedness Plan objectives are to: 1) Ensure the preparedness of operational services and other agencies for work during disasters; 2) Enhance the co-operation of various agencies in case of disasters; 3) Establish functioning organization coordinating rescue work in the injury area in case of disaster or a major accident; 4) Ensure the rescue of the maximum number of victims in case of disaster or major accidents; 5) Minimize the number of somatic and mental traumas; 6) On the basis of risk analysis, aim for the maximum reduction of the range of disaster consequences, and; 7) On the basis of risk analysis, draft and further update disaster medicine plans.

2. Structures Participating in Rescue Operations

The preparedness and disaster plan consists of five coordinated organizations to ensure that every aspect of response and rescue is performed efficiently in the case of a disaster. This structure includes all necessary tasks and services including rescue and response, law enforcement, medical response, border security, and national security.

The five organizations are: the Rescue Board, the Health Care Board, the Police Board, Border Guard Administration, and Estonian Defense Forces:

1. Rescue board directs, administers, and coordinates firefighting and rescue during a major accident, where necessary; coordinates large-scale operations or operations extending across the borders of several counties, and international operations; advises on development of crisis management plans

and present an opinion on the crisis management plans and the risk analyses to the crisis management committee.

2. Health Care Board administers the emergency preparedness of emergency medical service; organize responding to emergencies in co-operation with other agencies, regional hospitals and the Ministry of Social Affairs
3. Police Board coordinates the work of police prefectures, where necessary, in order to respond to emergencies.
4. Border Guard Administration coordinate search and rescue operations in case of air or sea accidents.
5. Estonian Defense Forces assist the rescue service with rescue operations, the police with security and siege operations, the EMS with administrating first aid and transportation [3].

3. Crisis Management at Regional Level

An advisory and cooperative body of the county government in an emergency will be the Crisis Management Committee. It consists of rescue service, law enforcement, and health. The county's rescue service will be responsible for defending lives, property, and the natural environment; saving lives and reducing the number of injuries; as well as fire fighting and rescue operations. The local police will ensure public safety and defend public order, search of missing persons, administer evacuations, and investigate activities. Finally, the Health Care Board will be responsible for providing emergency medical care.

4. Duties of Agencies Participating in Rescue Operations

4.1 Rescue Service

The Rescue Service will direct rescue operations; administer cooperation of services; administer communication between services and 112; specify the injury area and summon the crisis management team of the area.

In the injury area, the rescue service will be responsible for directing and performing rescue operations; protecting humans, property, and the environment against damaging factors; rendering lifesaving first aid to victims; establishing communications with services and SOS center; bringing the victims out of the risk zone; transmitting information; deciding on a need of evacuation.

The injury area the rescue service will specify:

- Management post
- Checkpoint(s)
- Casualties assembly point
- Treatment zone
- Assembly point for uninjured
- Assembly point for the dead and property
- Landing site for helicopters
- Decontamination point
- Risk zone
- Removal point

- Transit point

4.2 Police

The police will establish communications with other services, forward the incoming information of the accident promptly to the SOS center. The police officer on duty shall dispatch the senior police official on duty to the SOS center where the crisis management team shall gather.

In the injury area, the police will be responsible for establishing contact with rescue work coordinator; directing traffic, including landing of helicopters and regulating the traffic of rescue vehicles; documentation of the situation at the injury site; marking, securing, and guarding the injury site; evacuation of victims on the order of the rescue work coordinator; search of the injury area and its immediate surroundings; directing the uninjured persons involved in the accident to assembly point for uninjured for identification and registration; identification and registration of casualties; specification of the assembly point for the dead and administrating the transportation of the dead, identification and registration of them; performing investigation activities.

In the injury area the police shall specify and mark: the injury site; the assembly point for the uninjured; the assembly point for the dead and property.

4.3 Medical service (EMS)

EMS will procure extra medical equipment to the injury site; mediate information among the injury site and hospitals; organize appropriate vehicles for transportation of victims and transport the casualties; consult specialists where necessary; organize psychosocial assistance.

In the injury area, the medical services are responsible for:

- Specifying the range and character of the required medical care
- Administering prompt medical care to the injured, in case of chemical and nuclear accidents taking specific measures coordinating with the rescue service
- Examining victims and determining priorities of providing care
- Administering as fast transportation of victims from the injury area to the available medical institutions
- Supporting the vital functions up to the hospital arrival
- Steady cooperation with the services participating in the elimination of the accident
- Gathering information falling within its sphere of responsibility and forwarding it to the rescue work coordinator.

4.4 Medical Care Administration

The rescue work coordinator, forming the management post together with police official and medical officer, directs all rescue operations in the injury area. The rescue coordinator will appoint an injury site rescue coordinator and the chiefs of sectors, where necessary. The injury site doctor is the senior member of the first ambulance that arrives on the scene.

The doctor's responsibilities include: 1) Inspecting the injury site—verifying the number of the injured, characterize injuries, etc. 2) Dispatch on-site ambulance crews into action; 3) Gather information and forward it to the medical officer (in case of the management post missing s/he shall forward the information to the S.O.S center; 4) Responsible for request additional resources; 5) Receive/request from the medical officer the information regarding the location of treatment zone; 6) Appoint the treatment zone doctor if the treatment zone has been established, and direct a part of the medics into treatment zone; 7) Cooperate with the injury site rescue work coordinator; 8) Be the last person to leave the injury site after getting permission from the rescue coordinator.

If the management post has been formed, a medical officer shall be selected and appointed from the arriving ambulance crews.

The *medical officer* shall

- Cooperate with members of the management post
- Process the information arriving from the injury site
- Administer traffic of medical staff and medical equipment
- Ensure optimal usage of resources
- Intermediate information between the SOS center and the injury site: forward the information about the number of the injured; forward the necessary information of injuries to the hospitals via the SOS center; request the additional assistance, free capacity in the hospitals etc; forward to the removal point medic the information of the hospitals admitting the victims.

The members of arriving ambulance crews shall operate in the injury area: under the subordination of the injury site medic they perform first triage; administer lifesaving first aid; ensure the organized traffic of victims pursuant to the scheme or shall operate in the treatment zone: perform secondary triage; perform medical treatment procedures under direction of treatment zone doctor or shall transport victims to the hospital follow instructions of removal point medic.

The Injury site doctor shall appoint the treatment zone doctor. After reporting to medical officer *treatment zone doctor* shall

- Perform or direct secondary triage
- Decide the range of medical treatment possible
- Request additional equipment from the medical officer where necessary
- Appoint the removal point medic
- Direct victims prepared for transportation to the removal point

The *removal point medic* shall

- Establish communications with the management post, receiving information regarding hospitals admitting victims and transportation opportunities
- Organize transportation of victims to the most appropriate hospital
- Note the name of the receiving hospital on the triage card
- Cooperate with the treatment zone doctor and the removal point police

The call signs of EMS shall be changed during disasters in the injury area as follows: leaders of ambulance crews shall report to the SOS center via radio of taking a leading role at the injury site (e.g. first arrived EMS team leader shall become the injury site doctor, notify the SOS center of his/her radio call sign change and receive the corresponding confirmation from the SOS center).

5. The Tactics of Aiding the Victims at the Injury Site

All units involved in elimination of disaster consequences shall administer first aid aimed at saving the lives of as many casualties as possible. Recognizing the conditions endangering life and administering common first –aid procedures: positioning, opening the airways, and stopping bleeding – saves lives and everyone shall perform it.

Professional medics shall perform first triage at the injury site. During first triage respiration, blood circulation and state of consciousness shall be assessed and only minimal aid shall be given – lifesaving first aid. First triage shall be the basis for the order of transporting the victims to the casualty assembly point or the treatment zone.

Secondary triage shall be usually performed in the casualty's assembly point or the treatment zone, where victims shall be examined more thoroughly and the priorities determined during first triage shall be re-assessed. Secondary triage shall be the basis for the order of transporting the victims to the hospital.

The overall aim of the triage is to help as many casualties as possible and use the personnel and equipment most rationally. Triage cards shall indicate priorities for aiding victims.

6. Defense Forces

A request for the inclusion of the Defense Forces shall be submitted by the Disaster Relief Team of the Rescue Board, which will administer participation of units of the Defense Forces and the Defense League [3].

References

[1] State of Emergency Act 1996
[2] Regulation No. 222 of the Government of Estonia of 10.07 2002.
[3] Regulation No. 245 of the Government of Estonia of 26.09 2003.

© 2006 IOS Press. All rights reserved.

Preparing for Preparedness: Emergency and Disaster Medicine System in Latvia

M. BUSMANE, MD[a], M. SICS, MD[a], R. PUPELE, MD[a], E. AKITIS[a]
[a]Centre of Emergency and Disaster medicine Ministry of Health, Latvia
Pilsonu Str. 13 Riga, Lv-1002 Latvia, Maija.Busmane@kmc.gov.lv

Abstract Becoming a member of the European community and NATO opens many opportunities for Latvia, but at the same time we have to revise how prepared we are to prevent the consequences if emergency occurs. Concerning emergency and disaster medicine, former tools and organizational structures are not corresponding any more to the current situation. In this article, we introduce the current and upcoming structure and organization of the Emergency and Disaster Medicine System in Latvia. We have paid attention particularly to the essential component of Emergency and Disaster Medicine System- pre-hospital Emergency Medical Services. This new approach reveals better possibilities for the State's stewardship in this field and increases response capacities for emergency and disaster medicine management.

Keywords. Disaster preparedness, emergency, Latvia, public health

Introduction

During 2004–2005, several unexpected natural and man-made events disrupted Latvia's routine of life. In September 2004, an earthquake, measuring five on the Richter scale, frightened people throughout the country. Earthquakes in Latvia are rare and an entire generation can grow up without experiencing one. The next disaster followed in January 2005 when violent storm of hurricane-like proportions caused a power blackout of almost the entire State, causing huge damages to the public and private economy. Large floods and forest fires are more common in Latvia, with consequences that seem minimal compared to those that followed the blackout. At the turn of 2005, several serious traffic accidents with mass causalities happened almost simultaneously. Threats of terrorism are becoming a real concern due to our geographical proximity to countries where bombings are taking place. There is a big difference between the assessment and situational description, giving a sense of how safe everybody feels in his or her habitat, environment and living space, from the point of view of experts and the general population.

As Latvia becomes a member of the European community and NATO, many opportunities will be available. However, at the same time, we need to re-evaluate how prepared we are to prevent the consequences in case emergency occurs. "Crisis is a force for change"—this is the experience-based truth. Although we cannot say that Latvia went through a crisis, the problems of the past several years discussed above made the development of emergency and disaster medicine progress more rapidly.

When dealing with emergency and disaster medicine, we have to recognize that tools and organizational structures utilized in the past are no longer adequate. We must look for new structural solutions and a new coordination framework. The local national capability has become more and more relevant for the international society. Under these circumstances, the Concept of Preparedness and the process of its development obtains a new meaning; Therefore, specialists of our Centre of Emergency and Disaster medicine, in collaboration with other professionals and experts have performed the study.

1. Study Objectives and Methods

We received full cooperation among the health professionals and other rescue services providers that was necessary to obtain the best outcome from emergency and disaster management. In the course of this study, we paid special attention to the essential component of Emergency and Disaster Medicine System (EDMS)—pre-hospital Emergency Medical Services (p-h EMS) structure and management.

1.1. Objectives

The objectives for our study were to: 1) Analyse the current structure and functionality of p-h EMS; 2) Create the up to date model of p-h EMS; 3) Analyse what the term preparedness means for health care stakeholders, and how to provide it in the best possible way in the context with health care reforms, civil military crisis management, and international cooperation.

1.2. Methodology

The emergency profile is a critical concern of the health care system where it is often difficult to make decisions for change. We recognized the significance of sharing our experience with other countries, and used the lessons learned to avoid possible mistakes.

When we discuss Emergency and Disaster Medicine (EDMS) the term "disaster" refers to "a situation in which the need of acute medical care exceeds the immediately available resources, and in which extraordinary and coordinating measures are necessary if normal quality standards are to be maintained." Regarding the term "mass casualty," we refer to the demand on medical facilities that drains available resources to the point where even after reinforcements, normal care and the quality of care suffers, and might be limited.

In the exploration of the opportunities for collaborative work if a disaster or mass casualty occurs, we are evaluating the everyday emergency response results. During this study we used the following methods: data collection, site investigation, questioning, assessment, comparison, and forecasting.

2. Indicators of p-h EMS availability

Well-known essential indicators of p-h EMS availability are the sudden death rates from transport accidents and Ischemic heart disease.

Table 1. Sudden Death Rate per 100,000 of population, 2003 Source: European HFA Database, June 2005

Transport accidents		Ischemic heart disease	
Lithuania	23,76	Lithuania	327,75
Latvia	22,66	Estonia	312,53
Poland	15,5	Latvia	291,58
Hungary	14,7	Hungary	232,66
Estonia	13,64	Czech republic	176,09
Czech republic	13,33	Malta	149,77
Slovenia	13,19	Poland	124,93
Malta	3,5	Slovenia	94,37
Average EU	11,6	Average EU	105,76

When we compare the SDR from Ischaemic heart disease within Latvia during last 20 years (1983 – 423; 2003 – 291,580) we see that improvements have been achieved, but in comparison to other European countries, the mortality rates are still high. We cannot underestimate the role of preventive measures, but at the same time we have to admit that if patients would be delivered to specialists on time, the explosion of intravascular heart surgery in Latvia could reduce the sudden death rate from the Ischaemic heart disease remarkably. If death caused by traffic accidents are medically remedied, timeliness and the quality of p-h EMS has a direct correlation to the seriousness of the consequences—every lost minute reduces survival rate.

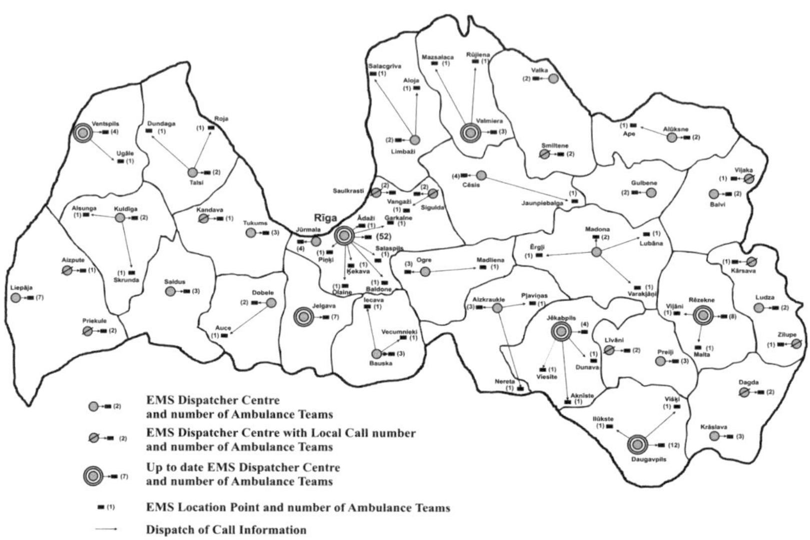

Figure 1. Location of p-h EMC Dispatcher centres with different technical abilities (Latvia, 2005)

3. Description of Pre-hospital Emergency Medical Services Today

The infrastructure of p-h EMS was developed as unified body where separate units were owned by municipalities with different capabilities. P-h-EMS institutions provide emergency help for about 2.3 million inhabitants in an area of 64,589 km [2]. Latvia has 26 local municipality districts and 444 parishes.

Legislation requires that in the case of disasters and mass casualties, all municipal p-h EMS institutions must be under the operational management of the State institution responsible for provision of EDMS—the Centre of Emergency and Disaster Medicine. The Centre is responsible for the coordination and management of EDMS planning, coordination and management of unitary knowledge qualification system, and operational management of EDMS. The Operative Medical Service unit provides specialized emergency medical care for the hospitals throughout the entire State in 24 hours, and supports the local p-h EMS in case of medical disaster. Despite the delegacy, there are many obstacles to carrying out this mission in the best possible way:

- Too many Dispatching Centers with low capacity, without interconnection
- Municipality border problems
- Unique EMS call problem
- Cost ineffectiveness
- Lack of Ambulance Teams (personnel)
- Irregular location of Ambulance Teams
- Lack of ambulances, technical and medical equipment shortages.

The overall financial resources for health care are too little, but additionally existing decentralization of p-h EMS infrastructure leads to cost ineffective resource management. As a result of prolonged decentralization of p-h EMS, we have a lack of ambulances, shortage of technical and medical equipment, and irregular location of Ambulance Teams. About 40 % of ambulance vehicles are outside of capital and are, on average, more than eight years old.

Direct call number for p-h EMS have existed for a long period of time, but due to the evolution of the communication system, previous technical solutions do not correspond to the requirements to address the modern day needs. When our people ask only for medical help, they rarely use the single European emergency call number, 112.

Table 2. P-h EMS calls and calls completion (daily average)

Area	**Calls received by p-h EMS Dispatcher's**		**Calls completed by Ambulance Teams**
	In all	Including through "112"	
The Capital Riga	1,000	13 %	500
Entire Country	2,500	11 %	1,400

Thirty-seven daily dispatcher centres with dissimilar technical abilities, and without interconnectivity, receive health emergency calls from people, deliver information to the Ambulance Teams, and collect the information after calls are completed. Movement of ambulance vehicles is restricted in certain areas (e.g., borders are municipalities` responsibilities). Only in the Capital of Riga and in several larger cities, Dispatcher Centres Business processes are modern and up-to-date. Difficult job conditions, emotional and physical stresses are not proportional with emergency professionals` salaries, which accounts for high labor turnover and lack of Ambulance

Teams personnel. Therefore, there is no capability to keep uniform EMS quality standards all over the State.

The problems mentioned above directly affected the timeliness and quality of p-h EMS. After data collection, site investigation, questioning, assessment and comparison, we made the following conclusions.

4. Study Conclusions and Subsequent Considerations

- There are no facilities for effective management of EDMS shared resources in circumstances while p-h EMS's providers management is decentralized;
- There are unequal possibilities to receive p-h EMS in day-to-day and especially if a disaster occurs;
- New infrastructure and organization for medical institutions involved to prevent emergency consequences is certainly necessary.

We have considered several principles and approaches for creating a new structure and organizational scheme of p-h EMC to cover the patients` needs for equal availability of services (on time and qualitative): subsidiary, coordination, succession, consolidation of resources, and inter-sector scope.

During the elaboration of the white paper we had to recognize that essential attention must be paid to all transitions of health care system and health professionals` attitude towards reforms. Under the Soviet system, all health care services were free and generally accessible to the entire population. Tremendous changes, which mainly concern primary health care and hospital care, have affected our health care system in last 15 years. Trends for health care providers to increase their entrepreneurial activity are a burden for the p-h EMS day-to-day workload. The pressure on hospitals to cut the time patients spend in the hospital, general practitioners' unwillingness to work beyond normal office hours, resistance from large part of health professionals` to assist in reforms, which are socially important but not so convenient for them, has led patients to look toward the p-h EMS for assistance. There is why EMS calls can be used for monitoring of health care services availability.

5. Description of the Pre-hospital Emergency Medical Services in the Near Future

After long discussions, assessment, and comparison with EMS organizational experiences in other countries, it was decided to upgrade the country's EMS ambulance vehicles, Ambulance Teams, and dispatcher centres. Unified and integrated dispatcher and management information system under centralized control will be introduced in order to ensure rapid, efficient, and equal access to p-h EMS for all people in Latvia.

5.1. Concept of the Project

Centralized and integrated p-h EMS management information has to ensure an efficient national coordination and collaboration with:

- different regions of the country;
- health sector bodies (hospitals, insurers, health services purchasers, etc.);
- emergency and rescue services (fire and rescue services, police, etc.);

- institutions and agencies involved in disaster management (Public Health Agency, Radiation Safety Centre etc.).

The aforementioned calls for organizational changes, changes in work-flow processes, improved equipment and technology in the dispatcher centers and in the ambulance vehicles, improved communication networks among units, and the development of a centralized management information system. It calls for new regulations and new skills as well. Besides, it requires equivalent knowledge, good will, and a "common" language between health professionals and health managers.

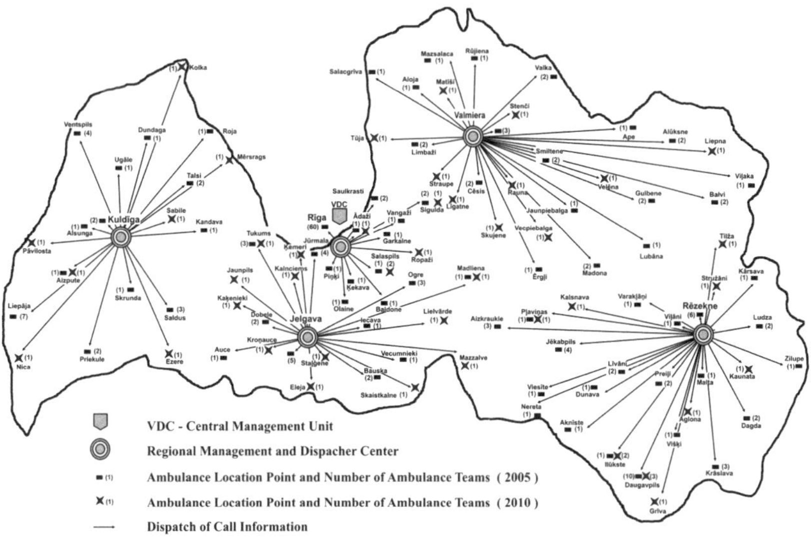

Figure 2. Location of p-h EMC Regional Dispatcher Centres with up to date technical abilities (Latvia, 2010)

It was foreseen that the future EDMS basic component of p-h EMS will be organized in three distinct levels comprising: (1) Central Management Unit (CMU) located at the Centre of Emergency and Disaster Medicine in Riga, (2) Five Regional Management and Dispatcher Centres (RMDCs), (3)Ambulance Location Points network and Ambulance Teams.

Central Management Unit (CMU)

The function of the CMU will be solely to monitor, supervise, and manage all regions of Latvia, i.e., the CMU will not handle emergency calls from the public. Only in the case of major disasters, will the CMU have an operational role in dispatching resources. The CMU as an infrastructure of the State institution responsible for EMDS, and will be the head office exercising command and control over five RMDCs. The workstations of CMU also could be used for various roles such as specialized emergency dispatcher, as a reserve for RMDCs workstations, and videoconference workstations.

Five Regional Management and Dispatcher Centres (RMDCs)

Each of the five RMDCs is planned for one functional region. The role of the RMDC will be to handle incoming emergency calls and dispatch the ambulances as well as being the primary response unit in case of a medical disaster. The RMDC will also be responsible for the regional coordination and collaboration with other health sector bodies and operative services (e.g., health insurers, hospitals, fire and rescue services, police, etc.).

The old principle that after receiving a call the nearest Ambulance Team goes to the scene of the accident will be determinative, without reference to municipality borders. Another important element concerns the routing of incoming calls to the RMDC, and calls between different RMDCs in case one RMDC fails. In such situation, all incoming traffic must be re-routed to an alternative RMDC.

Ambulance Location Points and Ambulance Teams

Ambulance vehicles and Ambulance Teams will be located in specially equipped places, under the supervision of the RMDCs. At these location points, Ambulance Teams will be charged to execute the actual mission. Every link of call completion business processes will be provided with up-to-date technologies.

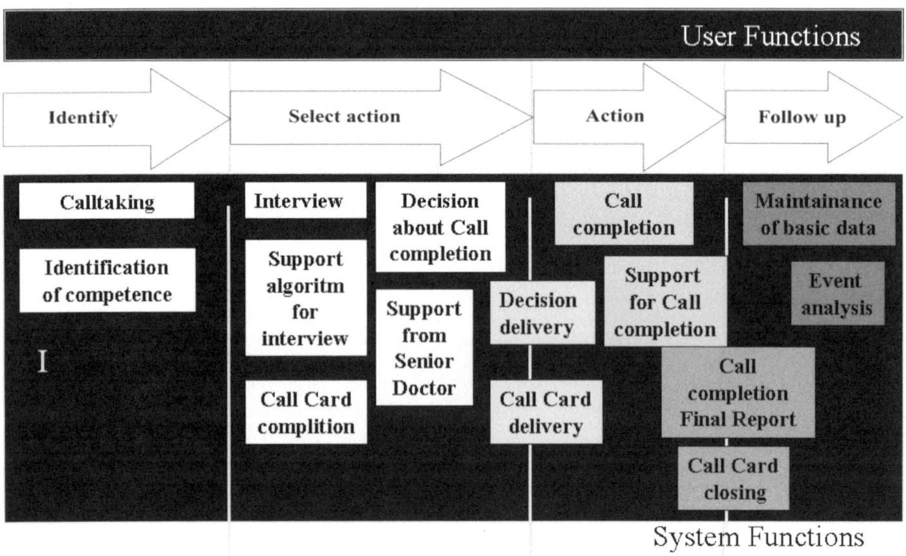

Figure 3. Call completion business processes

5.2. Implementation plans

The new p-h EMS Dispatch System will be carried out in a phased approach—the Pilot Study (1 RMDC + CMU), followed by national implementation. The Pilot Study will consist of three distinct parts:

- Main Technology: all hardware and software, e.g., servers, switches, PCs, radio and telephony systems and all application software for Computer-Telephony Integration, Geographic Information System, Computer-aided

Dispatching and mobile units. It also includes the Local Area Network infrastructure and communications, as well as set-up and configuration, testing, training, maintenance and support;

- Infrastructure: the communication (data and voice) between the RMDCs, the Ambulance Location Points, other health sector bodies, the CMU and various databases through a Wide Area Network (WAN); it also deals with the establishment of connection between the RMDC and the mobile units in the ambulance vehicles.
- Facilities: physical facilities for the CMU, RMDCs, and the Ambulance Location Points, as well as with installation, testing and training.

The Technical scope of Pilot Study is generally set up. More specific technical solutions will be expected from the participants (applicants), and then the purchasing procedure of technical order will come up.

To proceed with entire development of p-h EMS certain activities required:

1. Increase the number of Ambulance Teams.
2. Enlarge the Ambulance Location Points network.
3. Provide p-h EMS with mobile communication systems.
4. Introduce the direct p-h EMS call number with update functionality.
5. Introduce the unified management of p-h EMS Dispatch system.
6. Establish a new remuneration system for p-h EMS staff, based on graduation of positions and social assurance.
7. Develop necessary legislative acts.

6. Comparison of Pre-hospital Emergency Medical Services Resources

Table 3. Pre-hospital Emergency Medical Services resources: Now and after

	2005	**2010**
1. Dispatcher Centres	37	1 + 5
2. Ambulance Location Points	80	115
3. Ambulance Teams	192	240
4. EM professionals	1728	2560

7. Benefits Expected after Implementing New Pre-hospital Emergency Medical Services System

- Save more lives (traffic accidents take approximately more than 100 lives per year);
- Increase p-h EMS accessibility to meet international standards (response time, unitary call number, quality);
- Increase p-h EMS management role as a support for international cooperation, for Civil Military Crisis Management assistance especially in case of CBRN threats;
- Increase capacities for Control and Statistics;
- Ensure synergy effect and efficient use of available resources;
- Harmonize cooperation and coordination with other Rescue services;

- Assure coordination of specialized Emergency Care, Consultancy Service reserves, Hospital Services, Medical evacuation and Transportation Services and other Health resources;
- Increase capacity for Primary Health Care management (PHC central).

8. Updated Concept of Emergency and Disaster Medicine System in Latvia

During this study, we have reconsidered, complemented, and consolidated our concept of EDMS organizational structure and functionality—the Pyramid of Preparedness; we have also considered which items need to be covered in "peace time" in order to be prepared for the time a disaster (x) occurs.

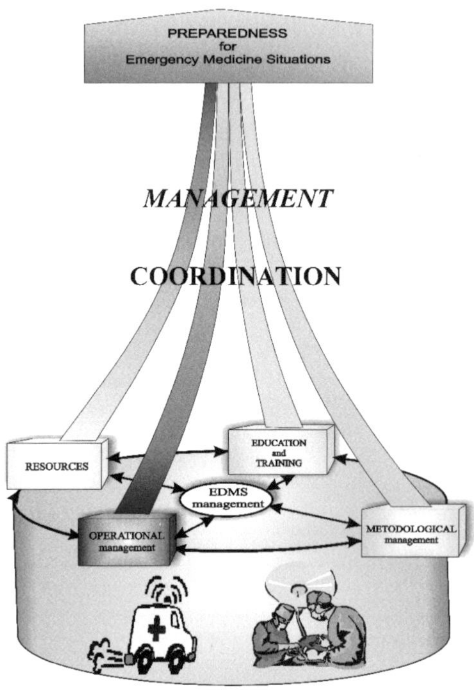

Figure 4. The Pyramid of Preparedness

For being prepared daily and in case of emergency several preconditions are relevant. The EDMS concept recognition allows us define the Pyramid of Preparedness and its cornerstones:

- Planning and methodological management
- Resource provision
- Education, training and international cooperation

- Monitoring of every day EMS and Operational management
- Highly developed hospitals` admissions and emergency units
- Highly developed p-h EMS, the basic operational unit of EDMS

The most important factor for preparedness, particularly if the resources are limited, is management and coordination. The recent events in the world and the risks assessment indicate the substantial role of training and exercising.

Revamping EDMS certainly costs a lot of money, but timely and reasonably carried out actions, speaking exclusively about traffic accidents, could save at least 100 lives per year. This is real power where the technology and politics can push all efforts together to extend the State economic productivity and well-being of their people.

After long and contentious discussionsThe Government of Latvia has now adopted the Policy Statement of EMS development 2005 –2010 (July, 2005).

The government delegates the lead role to maintain EDMS and its development process to the Centre of Emergency and Disaster medicine. It means that sustained control over socially valued activity conducted by a public institution will reveal the State's stewardship approach.While almost everyone would like to be the "boss" (no one likes to be coordinated and managed), especially in the case if health care providers who have become an independent entrepreneurs, the health care budget is permanently deficient, and the State must lead the way.

So, in light of the sweat and blood, more efforts will be required than ever before in order to enforce comprehensive plans and projects, which are not challenging provider's earnings increase, but are still extremely essential for safeguarding social objectives. Moving forward, research-based experiences, proper psychological approach, and strong control over all processes again will be on agenda of the State Centre of Emergency and Disaster medicine but this will be the next story. *To be continued . . .*

References

[1] Starling G. Managing The public sector (1993) By Harcourt, Inc, By Latvian School of Public Administration (1999).

[2] Hellenberg T. 2002. Challenging disasters. series A: 7 Helsinki Kikimora Publications: Aleksanderi Instiute.

[3] Saitman RB, Busse R, Mossialos E. 2002. Regulating entrepreneurial behaviour in European health care systems. The European Observatory on Health Care Systems, WHO Regional Office for Europe, Copenhagen, Denmark.

[4] McKee M, Healy J. 2002. Hospitals in a changing Europe. The European Observatory on Health Care Systems, WHO Regional Office for Europe, Copenhagen, Denmark.

[5] Report of the Commission on Macroeconomics and Health. 2001. World Health Organization, Geneva. Investing in Health for Economic Development.

[6] World Health Organization, Geneva 1999. ISBN 92 4154519 4, Community emergency preparedness: A manual for managers and policymakers.

[7] World Health Organization, 2004. Field manual for capacity assessment of health facilities in responding to emergencies.

[8] Unpublished materials of Centre of Emergency and Disaster medicine Ministry of Health, Latvia.

© 2006 IOS Press. All rights reserved.

Author Index